Hands-On Biology

Laboratories for Distance Learning

Mimi Bres
Prince George's Community College

Arnold Weisshaar
Prince George's Community College

Cassandra Moore-Crawford
Prince George's Community College

W. H. Freeman and Company • New York

Cover: Dana B. Bres. Brewer's Blackbird. The image was taken in Monterey, California. Brewer's are permanent residents in much of California and are found in open country in many Western states.

Photographs by Landon McMahon unless otherwise noted.

CHAPTER 4
Figure 4-1a, Dennis Kunkel/Alamy; **Figure 4-1b,** Michael Abbey/Photo Researchers; **Figure 4-1c,** Winton Patnode/Photo Researchers

CHAPTER 5
Digital stop watch, Jim Mills/PhotoXpress; **Figure 5-1,** Courtesy of Nasco

CHAPTER 6
Small pot, Dmitry Vereshchagin/Fotolia; **Ginger root,** Photodisc; **Funnel,** sanja gjenero/stock.xchng; **Figures 6-3, 6-4,** Courtesy of the author; **Figure 6-7,** Photodisc

CHAPTER 7
Figure 7-4a, Photodisc; **Figure 7-4b,** Biology Media/Photo Researchers; **Figures 7-5a-g, 7-6 through 7-9,** Courtesy of the author

CHAPTER 8
Figure 8-1, Visuals Unlimited/Corbis; **Figure 8-2,** Biophoto Associates/Photo Researchers 8; **Figure 8-5,** L. Willatt/East Anglian Regional Genetics Service/Photo Researchers

CHAPTER 10
Figure 10-6, AMRESCO, Inc.

CHAPTER 11
Digital stop watch, Jim Mills/PhotoXpress; **Figure 11-1,** Futch, S.H., et. al. A Guide to Scale Insect Identification. Publication #HS-817. Figure 7. UF/IFAS; **Figure 11-3 (top row: L-R),** Lee Karney/U.S. Fish and Wildlife Service; EyeWire, Inc.; Dr. Thomas G. Barnes/U.S. Fish and Wildlife Service. **(middle row: L-R),** Photodisc; Photodisc/Punchstock; Courtesy of the author. **(bottom row: L-R),** EyeWire, Inc.; Courtesy of the author; Fjuditu/Morgue File. **Figure 11-4 (first row: L-R),** Courtesy of the author; ril/photoexpress; degrees/Stock.xchng; katmystiry/Morgue File. **(second row, L-R),** ImageState; courtesy of the author; Alamy; kerry/photoxpress. **(third row, L-R),** rockphoto/Featurepics; wedhatted/Morgue File; clconroy/Morgue File; courtesy of the author. **(fourth row, L-R),** Corbis; Photodisc/Getty Images; xandert/Morgue File; Photodisc. **Figure 11-5 (left, middle),** Courtesy of Juan M. Hurlé; **(right)** Lena Andersson/Dreamstime.com. **Figure 11-6 (left, middle),** Courtesy of Juan M. Hurlé; **(right)** Ingo Arndt/naturepl.com. **Figure 11-7 (left),** Oxford Scientific/Photolibrary; **(right)** Photodisc/Getty Images

CHAPTER 12
Digital stop watch, Jim Mills/PhotoXpress

CHAPTER 13
Digital stop watch, Jim Mills/PhotoXpress; **Figure 13-1 through 13-3, 13-5,** Courtesy of the author; **Figure 13-6,** Laboratory of Tree-Ring Research, University of Arizona

CHAPTER 14
Gallon of water, Fotomak/Dreamstime.com; **Bucket,** Agtha Brown/Stock.xchng; **Digital thermometer,** DreamEmotion/PhotoXpress; **Digital stop watch,** Jim Mills/PhotoXpress; **Blue and red pencil,** Artville

CHAPTER 15
Figures 15-1, 15-2, 15-4, Courtesy of the author

CHAPTER 16
Figure 16-1, Courtesy of the author; **Figure 16-2,** George Doyle/age fotostock; **Figure 16-3,** Courtesy of the author

CHAPTER 17
Figure 17-1, AFP/Getty Images

CHAPTER 18
Digital stop watch, Jim Mills/PhotoXpress

CHAPTER 19
Figures 19-1, 19-2, Courtesy of the author

APPENDIX II
plant; brown leaf, Courtesy of the author

APPENDIX IV
pH paper, ©Carolina Biological Supply Company, Used by permission; **Glucose test strip,** ©Carolina Biological Supply Company, Used by permission; **Lactaid,** Will & Deni McIntyre/Photo Researchers; **Protein test strip,** Courtesy Science Kits; **Pipet,** ©Carolina Biological Supply Company, Used by permission; **Limewater, sealed tube,** Courtesy Science Kits; **Strip thermometer, wide range,** Courtesy Science Kits; **Strip thermometer, narrow range,** Courtesy Science Kits; **Dialysis tubing,** Courtesy of Nasco; **Stamp pad,** Photosindia/Photolibrary; **Magnifying glass,** brokenarts/stock.xchng

© 2011 by W. H. Freeman and Company

ISBN-13: 978-1-4292-5749-7
ISBN-10: 1-4292-5749-0

All rights reserved

Printed in the United States of America

First printing

W. H. Freeman and Company
41 Madison Avenue
New York, NY 10010
Houndmills, Basingstoke
RG21 6XS England

www.whfreeman.com

Contents

Exercise 1	Scientific Method	1
Exercise 2	Introduction to Chemistry	11
Exercise 3	Organic Molecules and Enzyme Action	25
Exercise 4	Cells and Cell Processes	35
Exercise 5	Diffusion and Osmosis	45
Exercise 6	Photosynthesis and Cell Respiration	57
Exercise 7	Mitosis and Meiosis	69
Exercise 8	Human Genetics	89
Exercise 9	Molecular Genetics	103
Exercise 10	Biology in Forensic Investigations	113
Exercise 11	Evolution	135
Exercise 12	Determining Your Ecological Footprint	157
Exercise 13	Growth Patterns and Nutrient Transport in Plants	169
Exercise 14	Homeostasis and The Circulatory System	187
Exercise 15	Studying Organ Systems Through Dissection I	199
Exercise 16	Studying Organ Systems Through Dissection II	209
Exercise 17	Metabolism and Nutrition	219
Exercise 18	The Nervous System	225
Exercise 19	The Reproductive System	237
Exercise 20	The Immune System	249
Appendix I	Making a Graph in Microsoft Word 2003	265
Appendix II	Starch Test Results	270
Appendix III	Using the Random Number Generator in Microsoft Excel	271
Appendix IV	Photo Index of Your Lab Kit Supplies	274

Preface

More and more colleges and universities are offering courses and even entire degrees online. The difficulty in providing online experiences in the sciences has been how to design lab activities so that experiments could be performed in a distance learning setting.

Hands-On Biology was developed for a non-majors introductory biology course with one specific goal in mind—to provide you with a parallel experience to what you would have if you were actually on campus taking a lab course. The themes selected for inclusion not only coordinate closely with lecture topics, but have specific relevance to student's lives and their roles in society. The activities address topics of current interest, such as nutrition, health, genetics, and environmental issues.

The format of the lab exercises recognizes that distance learning students have complicated work and travel schedules. You can take the book with you and complete your lab activities in other locations, even those without Internet access. The supplies needed are easy to obtain and the experiments can be performed in hotel rooms, on ships at sea, and in other remote locations.

Specific instructions are provided so you can complete the activities successfully, even if help from an instructor is not immediately available. Each exercise is broken into separate activities that are related to a particular topic. It isn't necessary to complete all the activities in one sitting. You can break them up to fit your own time frame.

Even though you're not in a campus laboratory, doing the experiments will allow you to be an active participant in your own learning. We hope the topics will be entertaining as well as informative and that you'll develop a greater appreciation for biology and its importance to our daily lives.

Best wishes,

Mimi Bres, Arnold Weisshaar, and Cassandra Moore-Crawford

Acknowledgments

Many thanks to:

Christine Barrow, Ph.D., Dean, Division of Science, Technology, Engineering, and Mathematics, Prince George's Community College for lending her expertise to the development of activities on molecular genetics and evolution.

The Biology 1010 students at Prince George's Community College (both face-to-face and online) for helping us thoroughly test these activities.

The biology faculty "lunch bunch" for providing ideas and moral support throughout the development process.

REVIEWERS

The reviewers listed below suggested many useful changes. Thanks for your assistance.

Dani Ducharme, Waubonsee CC

Kelly Burke, College of Canyons

David Eberiel, UMass Lowell

Fran Norflus, Clayton State

David Byres, Florida CC

Robin Gibson Brown, East Carolina University

Warren Hunicutt, St Pete College

Frederick Hayes, St Pete College

Pat O'Mahoney-Damon, University of Southern Maine

Anne Lumsden, Florida State University

Eddie Lunsford, Southwestern Community College

Sandra Tedder, Northwest Arkansas CC

Charlease Kelly, Claflin

Louise Nolan, Middlesex Community College

Jamey Thompson, Hudson Valley Community College

Karen Kettler, West Liberty University

John Kell, Radford University

Exercise 1 • Scientific Method

OBJECTIVES

After completing this exercise, you should be able to:

- formulate a testable hypothesis
- explain what a control is and why it validates experimental results
- collect and organize data for analysis
- calculate the mean, median, mode, and range for a set of data
- construct graphs that clearly and accurately represent a set of data points
- identify the characteristics of well-constructed versus poorly constructed graphs

SUPPLIES

Activity 3

SUPPLIES FROM LAB KIT

- none needed

HOUSEHOLD SUPPLIES

metric ruler, 30 cm (12 inch), 1

ACTIVITY 1 • STEPS OF THE SCIENTIFIC METHOD

When scientists approach a question or a problem, they do it in a very simple and structured way called the scientific method. Even though you're not a scientist, you probably use the steps of the scientific method frequently to solve problems that occur in your daily life. In general, the process that scientists (and everyday citizens) use to answer questions includes the following steps:

Making observations

If you go outside and your car won't start, that's an observation. You might make additional observations when you seek more information by checking the owner's manual or visiting the manufacturer's Web site.

Developing hypotheses

A hypothesis is a probable explanation based on observations. A hypothesis **should be based on evidence,** even if that evidence is your own prior experience. It must be logical. A good hypothesis has two essential features: it's **specific** and it's **testable.** The more specifically the hypothesis is stated, the easier it is to develop a method to confirm or reject the hypothesis.

Hypotheses are used to form predictions about the outcomes—the results that you would expect if your hypothesis is correct. Referring back to the car problem (your car won't start), you might hypothesize that the battery is dead or that the car doesn't have any gas. A hypothesis can lead you to make a prediction, similar to the following: "If the car battery is dead, charging the battery will allow the car to start."

Testing hypotheses with experiments

An experiment, in its simplest form, is a test that can confirm or falsify your hypothesis. What tests could be performed to determine whether your battery hypothesis is valid? One possibility is to test the charge with a battery tester or use jumper cables to provide a boost to start the car.

Collect data and analyze results

Based on the results of your experiments, you can determine whether the evidence supports your original hypothesis. Once a conclusion is reached, that's not the final answer. Conclusions are always subject to review and interpretation as more information is collected about the topic.

If the results don't support the original hypothesis, scientists usually form new hypotheses and try again. For example, if you try to jump start your car but it doesn't work, you'll need to form a new hypothesis about the cause of the problem.

INQUIRY AND ANALYSIS

1. To gain some experience in recognizing and composing well-framed hypotheses, **mark** the poorly framed hypotheses in the following list. **Rewrite** each hypothesis that you marked into an acceptable format.

   Reminder: Hypotheses must be clearly written, specific, and testable.

   a. _____ Spinach makes you stronger.

   b. _____ Adding fertilizer to tomato plants will result in higher fruit production.

   c. _____ Susan will lose weight by taking Quick-Loss Diet Pills.

   d. _____ Soft contact lenses are better than hard contact lenses.

   e. _____ Eating trans fat is bad for you.

2. Outline a situation you've encountered in your work or daily life in which you used the scientific method to solve a problem. Explain how you approached this problem by using the steps of the scientific method discussed in the beginning of this activity.

ACTIVITY 2 • DETAILS OF AN EXPERIMENTAL SET-UP

Although many types of research methods exist in science, most biologists regard a controlled experiment as the method of choice. A control provides a baseline for comparison with your experimental results, as shown in the example below.

> I looked into my neighbor's yard and noticed that his rose bushes are big, bushy, and covered with flowers. I said to him, "How did your roses grow so big?" He replied that he fertilized the bushes. I said, "Great—I can do the same to make my bushes as wonderful as yours. By the way, how can you be sure that the fertilizer was the cause of all this growth?"
> And then he said, "Gosh, I really don't know for sure. I fertilized them and they grew pretty well this year. I guess that the fertilizer caused this excellent growth." And I replied, "Are you sure it wasn't because we had more rainfall this year? Or that the weather wasn't as hot as usual?"

So, we're really not sure the fertilizer **was** responsible for the excellent growth.

INQUIRY AND ANALYSIS

1. List three factors that could have affected the growth of the rose bushes, in addition to fertilizer, rainfall, and temperature.

Since there are many possible factors that could have affected the growth of the roses, it would be useful if we could compare my neighbor's results with a **control** group. A control is a group of test subjects (in this case, rose bushes) that are the same species, live in the exact same conditions, receive the same amount of sunlight, get the same amount of water, everything the same. The only difference between the two groups is that one group gets fertilizer and the other doesn't. That way if there **is** a measurable growth difference between the two groups, you have more confidence that the result was due to fertilizer because there was no other difference between the two groups of rose bushes.

The same thing would hold true if you were testing a new blood pressure medication. You need a **control group** as a standard for comparison to make sure that your results are due specifically to the factor for which you're trying to test. The factor being investigated in a particular experiment is called an **experimental variable.**

In the case of a medical experiment, it's really hard to make the control group exactly the same as the experimental group. For example, if you give the **experimental group** medication in pill form, the control group has to take an identical pill (without the medication).

It's actually quite difficult to design the fake medications (called **placebos**). For example, if the real medicine tastes bitter, the placebo pill has to be bitter as well, so no one will know whether they're getting the real medicine or not.

Consider the following experiment:

A vitamin company wanted to determine whether large doses of vitamin C would shorten the length of time it takes to recover from a cold. The company hired 300 people and split them into two groups. The distribution of people in each group was similar with regard to age, weight, height, gender, etc.
   For the first ten days of the experiment, Group A was given a daily dose of 4 g of vitamin C (in a pill). Over the same ten days, Group B was given the same type of pill, but with no vitamin C content.
   None of the people in either group knew which pill they were receiving. In addition, the scientists analyzing the results of the experiment didn't know which people had been assigned to Groups A and B.
   On day one of the experiment, both groups were exposed to the cold virus by nasal swab. Researchers tracked the patients in both groups for the presence of cold symptoms.

In regard to the vitamin C experiment:

2. What variable was being tested?

3. Which was the control group?

4. Which was the experimental group?

5. Why did the researchers try to make sure the people in the two groups were similar in age, weight, health, etc.?

6. Why was it important that neither the test subjects nor the scientists analyzing the results knew which people actually took the vitamin C?

Biotechnology Today

Biotechnology is a new and growing field where biology, engineering, chemistry, and several other sciences come together. Biotech companies develop techniques that enhance and improve areas of medicine, control of diseases, waste management, energy production, food safety, agriculture and much more. New biotech applications are made possible through use of the scientific method.

Look for the icon in each chapter of this book to learn how biotech discoveries affect your daily life.

ACTIVITY 3 • COLLECTING DATA AND ANALYZING THE RESULTS

Now that you're familiar with the basics of an experimental set-up, it would be helpful to practice collecting data.

When your hand is spread wide, the **distance from the tip of your thumb to the tip of your little finger** is called your **hand span.** Hand span differs among individuals. In this activity, you'll be collecting hand span measurements and looking for patterns in the results.

INQUIRY AND ANALYSIS

1. Using a metric ruler, measure the span of your **right hand** as shown in **Figure 1-1**.

   Note: Round off your measurement to the nearest centimeter.

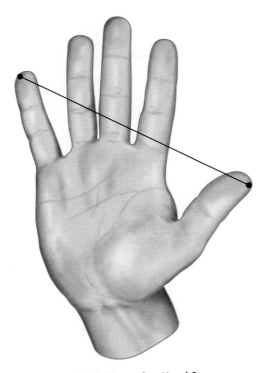

FIGURE 1-1 Measuring Hand Span

2. **Record** the results in the appropriate column of **Table 1-1**.

TABLE 1-1 Hand Span Measurements

Hand Span (cm) - Males	Hand Span (cm) - Females

3. Based on your hand span measurement (your observation), form a hypothesis about whether your hand span measurement is typical for an adult of your gender.

Hypothesis:

4. Measure the hand spans of **five additional adults,** to produce a total of **three** measurements for **males** and **three** for **females.**

   Note: Round off your measurements to the nearest centimeter.

5. **Post your results** as directed by your instructor.

6. Tabulate the pooled class results in **Table 1-2**.

TABLE 1-2 Pooled Class Results

Hand Span (cm)	Number of Males with Each Hand Span	Number of Females with Each Hand Span
16		
17		
18		
19		
20		
21		
22		
23		
24		
25		
26		
27		
28		
29		
30		
other		
TOTALS		

Within a set of numbers, the number that occurs most frequently is referred to as the **mode.** Knowing the mode can be useful in many situations. For example, if you were the buyer for a large shoe store, knowing the mode would help you decide what quantity of a specific shoe size you should buy to meet customer demand.

Considering the pooled results for the class, answer the following questions.

7. What is the mode for hand span among **males**? _____ cm

What is the mode for hand span among **females**? _____ cm

8. Subtract the smallest male hand span from the largest male hand span.

The difference in **male** hand span is _____ cm.

The difference between the largest and smallest numbers in a set is referred to as the **range.** This is a measure of how diverse (or spread out) the numbers are. For example, if there was a wide range of housing prices in your area, this could indicate that people with a variety of incomes live there.

9. Calculate the range of **female** hand spans. The range is _____ cm.

Does the range differ from your calculations for **males**? If so, how is it different?

10. To calculate the average hand span, also called the **mean**, add up all the hand span measurements collected for one gender. Then **divide** by the **total** number of people of that gender that were measured.

mean hand span for **males** _____ cm

mean hand span for **females** _____ cm

11. Is there a difference between the average male and female hand span? If so, describe the difference.

12. Was your original hypothesis supported? In other words, is **your** hand span typical for your gender? _____

If not, is your hand span smaller or larger than average? _____

ACTIVITY 3 • COLLECTING DATA AND ANALYZING THE RESULTS

13. Write out all the male hand span measurements in order from smallest to largest. If more than one person had the same hand span, **record that number multiple times.** Your list of numbers will look something like this:

15   16   16   17   17   17   18   19   20   20   20   20 etc.

This set of numbers is needed to determine the **median.** The median is the **middle number** of a set. When a set of numbers has a large range, the median often provides more accurate information than the mean.

 Note: If there are **two numbers in the middle** of your set (e.g., if your set has ten numbers), **average the middle two numbers** to determine the median.

Based on the pooled data, **calculate the median** hand span measurement for **males:** _____ cm

The median is often used when the data set includes a few extremely high or low values that would shift the mean away from what normally might be expected. The median is commonly used to represent home prices or income because the range of both home prices and salaries within a community can be extreme. For example, the presence of a well-known sports figure in your community would skew the mean salary calculation to the high end—well beyond that of the average citizen.

ACTIVITY 4 • GRAPHING YOUR DATA

A graph is a visual way of presenting your data. The purpose of a graph is to make it easier for someone to understand the information you want to express. Since graphs make it easier to present a lot of information in a small space, graphs are used in every aspect of society. It's important, therefore, that you're able to read a graph and interpret the data accurately. You should also be able to recognize the characteristics of a good graph and incorporate them into graphs of your own.

A graph has an **X (horizontal) axis** and a **Y (vertical) axis** (see **Figure 1-2**).

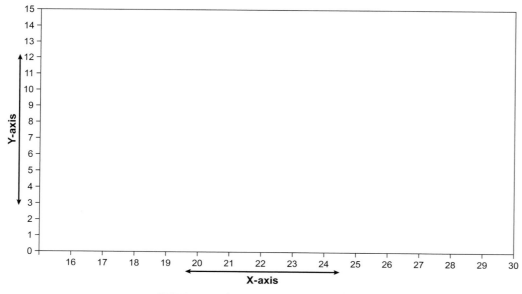

FIGURE 1-2 The X and Y Axes of a Graph

A good graph shows the **units of measurement**, an **explanation** for **both axes**, and has a **title** that lets the reader know what the graph is about. In addition, when the points are plotted, they should be **distributed along most of the X and Y axes**. Both axes should begin and end with numbers that are pretty close to the smallest and largest data points. To see an example of this type of data distribution, examine the distribution of the graphed points in **Figure 1-3**.

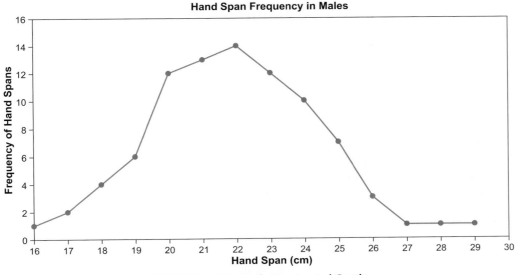

FIGURE 1-3 A Properly Constructed Graph

INQUIRY AND ANALYSIS

1. How many data points were plotted in **Figure 1-3**? _____

2. Which size hand span represents the **mode**? _____

3. What is the smallest hand span that was measured? _____

   What is the largest? _____

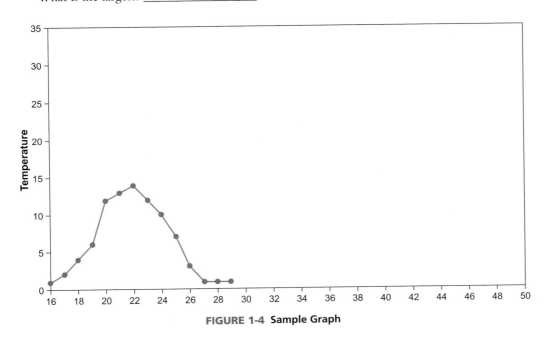

FIGURE 1-4 Sample Graph

4. List **three** items that should be corrected to improve the **presentation of the information in Figure 1-4.** Explain your reasoning for each correction.

   a.

   b.

   c.

5. **Create** a graph showing the data the class collected during the hand span investigation. Plot the hand span frequencies for both males and females on the same graph.

   Prepare the graph on a computer using the instructions in **Appendix I.** Submit the graph as required by your instructor.

Exercise 2 • Introduction to Chemistry

OBJECTIVES

After completing this exercise, you should be able to:

- diagram the basic structure of atoms, showing the number of protons in the nucleus and the number of electrons in each electron shell
- explain the octet rule and its relationship to the number of electrons needed for an atom's outermost electron shell to be stable
- compare and contrast ionic and covalent bonds
- distinguish between a polar and nonpolar covalent bond
- define the following terms: **acid, base,** and **pH scale**
- explain the formation of polymers by the process of dehydration synthesis and their breakdown to monomers by the process of hydrolysis

SUPPLIES

Activity 3

SUPPLIES FROM LAB KIT

- none needed

HOUSEHOLD SUPPLIES

food coloring or soy sauce, 1 tsp

tap water, 1/2 cup

vegetable oil, 2 TBS

measuring spoons, 1 set

measuring cups, 1 set

Activity 4

SUPPLIES FROM LAB KIT

- pH test paper, 10 strips

HOUSEHOLD SUPPLIES

cola, 1 TBS

vinegar, 1/2 cup

antacid tablets, 1 tablet

milk, 1 TBS

liquid detergent or
dish soap, 1 TBS

four other test liquids
(your choice), 1 TBS each

measuring cups,
1 set

small containers for
test liquids, 8

small bowl,
1/2 cup volume, 1

Activity 5

SUPPLIES FROM LAB KIT

- none needed

HOUSEHOLD SUPPLIES

paper clips, 7

Activity 6

SUPPLIES FROM LAB KIT

- glucose test strips, 2 strips
- Lactaid® tablets, 1 tablet

HOUSEHOLD SUPPLIES

milk,
1/4 cup

spoon,
1

ACTIVITY 1 • ATOMIC STRUCTURE

Elements are substances composed of many identical **atoms.** Atoms, in turn, are composed of **three types of subatomic particles: protons, neutrons, and electrons.**

The number of protons in the atom's nucleus is called the **atomic number.** It's the number of protons that makes atoms different from each other. Each type of atom has a different number of protons and, therefore, a different atomic number. For example, hydrogen has an atomic number of 1. This means that each atom of hydrogen has only one proton.

The names of elements are abbreviated with letters of the alphabet. For most atoms, the abbreviation is simply the first letter of the name (carbon is C, nitrogen is N, oxygen is O). Since the letter C has already been used for carbon, we have to use a different abbreviation for other elements beginning with C. In this case, the abbreviation has two letters. For example, calcium is abbreviated Ca, cobalt is Co, and chlorine is Cl.

Neutrons have no electrical charge, so they don't affect the overall charge of an atom. The number of neutrons in an atom may or may not be equal to the number of protons. Atoms with the same number of protons but a different number of neutrons are called **isotopes.** Most isotopes are unstable (**radioactive**) and decay over time. Radioactive isotopes have a wide variety of commercial and medical applications.

The number of electrons in an atom is equal to the number of protons. Protons have a **positive** charge and **electrons** have a **negative** charge. Since the number of protons (+ charges) is equal to the number of electrons (− charges), the atom as a whole is electrically neutral.

Electrons are arranged into layers around the nucleus of the atom. The layers are called **electron shells.** The **first electron shell** can hold a maximum of **two electrons.** The **second shell** can hold up to **eight electrons.** The **third shell** can hold up to **18 electrons.** Most of the atoms that are important in body chemistry will have no more than three shells, so the rules governing larger atoms aren't discussed here.

When the **outermost electron shell** of an atom is completely full with electrons, that atom is **stable.** Stable means that this atom **will not combine chemically** with other atoms. There's another way that an atom can become stable. If the **outermost electron shell** of an atom holds **exactly eight electrons,** the atom will be stable. This is known as the **octet rule.** When atoms combine together into molecules, they exchange or share electrons, and in the process, satisfy the octet rule.

INQUIRY AND ANALYSIS

1. For the following atoms, fill in the blank with the number of protons in the nucleus.

   Carbon (atomic number 6) has _____ protons.

   Sodium (atomic number 11) has _____ protons.

   Oxygen (atomic number 8) has _____ protons.

   Chlorine (atomic number 17) has _____ protons.

2. The **atomic number** of neon is **10**. Complete the diagram in **Figure 2-1** by inserting the **number of protons in the nucleus. Add dots** for the appropriate **number of electrons in each shell** (represented by the circles drawn around the nucleus).

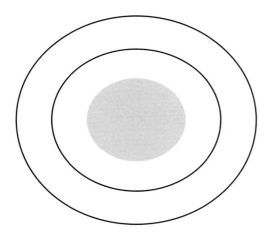

FIGURE 2-1 Atomic Structure of Neon

As you can see from your drawing, **all of neon's electron shells are holding the maximum number of electrons possible for each shell.** Since neon atoms are **stable** and don't participate in chemical reactions, they're safe to use in situations where a spark could cause an explosion (such as neon signs for restaurants and stores).

3. In **Figure 2-2,** complete an atomic diagram for chlorine. The **atomic number** of chlorine is **17**.

    Fill in the diagram with the **number of protons in the nucleus** and **add dots** for the appropriate **number of electrons in each shell.**

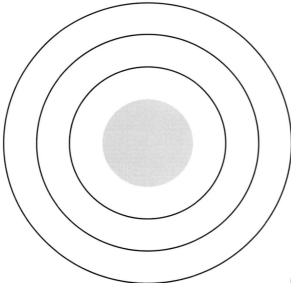

FIGURE 2-2 Atomic Structure of Chlorine

Since only the outermost shell of an atom determines how it will react chemically with other atoms, it usually isn't necessary to draw a full diagram of the atomic structure. Instead you can use a simpler drawing, called a **dot model**. A dot model shows **only** the number of electrons in the last shell.

In a dot model, the nucleus is represented by the chemical abbreviation for the atom and the electrons are shown as dots distributed around the abbreviation.

4. Sulfur has an atomic number of 16. Therefore, sulfur has _____ protons and a total of _____ electrons.

   The first shell would contain _____ electrons. The second shell has _____ electrons, and the third shell has _____ electrons.

   A dot model of sulfur, therefore, would look like the diagram in **Figure 2-3**.

FIGURE 2-3 Dot Model of Sulfur

   Does the sulfur atom satisfy the octet rule? _____ Is sulfur stable? _____

5. Based on your **completed diagram of the chlorine atom** in **Figure 2-2,** convert the atomic structure of chlorine into a dot model and draw it below.

As you can see from your drawing, the outermost electron shell of chlorine is not completely filled. Therefore, this atom is **not stable.**

To satisfy the **octet rule** chlorine needs an **additional** _____ electrons(s) in its outermost shell.

 Reminder: Stable means the outer shell is **completely full** or has **exactly** eight electrons.

ACTIVITY 1 • ATOMIC STRUCTURE **15**

Biotechnology Today

Radioactive isotopes are used throughout the medical field. One important application is bone imaging to locate stress fractures, a common injury of runners. To diagnose a stress facture, a patient is injected with 99Tcm, an isotope of technetium (atomic number 43), which has a half-life of six hours.

Technetium has a special affinity for bone tissue. Injured bone is more metabolically active and will absorb more of the isotope. As the technetium breaks down, it gives off gamma rays, which leave an image on the scanner screen. Since there's a higher concentration of 99Tcm at the site of the injury, it will be more visible on the resulting image.

ACTIVITY 2 • IONIC BONDING

It's possible for an atom to "pick up" an electron from another atom and satisfy its outermost shell by forming a chemical bond. When **electrons are donated by one atom and received by another,** this is **ionic bonding.**

The following is an example of **ionic bonding between atoms of sodium and chlorine.**

INQUIRY AND ANALYSIS

1. Sodium has the atomic number 11. In **Figure 2-4,** draw **two dot models** next to each other, one dot model of sodium (Na) and the other of chlorine (Cl).

Dot Model of Sodium Atom   Dot Model of Chlorine Atom

FIGURE 2-4 Dot Models of Sodium and Chlorine

As your models show, one of these atoms has a nearly filled outer shell, while the other is almost empty.

2. Circle the dot representing the single electron in the outermost shell of sodium. **Draw an arrow transferring that dot** to the outermost shell of the chlorine atom.

After the transfer is accomplished, has chlorine satisfied the octet rule? _____ **Explain your answer.**

16 EXERCISE 2 • INTRODUCTION TO CHEMISTRY

3. Now consider the sodium. Draw the dot model again, in **Figure 2-5,** this time with one electron missing.

FIGURE 2-5 Dot Model of a Sodium Atom after the Transfer of one Electron to Chlorine

4. Has sodium now satisfied the octet rule? _____ **Explain your answer.**

5. After electrons have been added or removed from an atom, **the number of positively and negatively charged particles is NO LONGER EQUAL.** Sodium has released one electron, which means that one negative charge has been removed. Chlorine has accepted an additional electron, which means that one negative charge has been added.

   (Circle one answer.) Before the electron transfer, sodium was **positively charged/ negatively charged/neutral.**

   After the transfer, sodium has _____ positive charges and _____ negative charges.

   After the transfer, sodium has _____ more positive charge(s) than negative charges.

   (Circle one answer.) So, the sodium now has a **positive/negative** overall charge.

6. Since we know that atoms are neutral in electrical charge, and we see that sodium is no longer neutral, the resulting atom is referred to as a **sodium ion.**

   (Circle one answer.) Sodium has formed a **positive/negative** ion.

   (Circle one answer.) Before the electron transfer, chlorine was **positively charged/ negatively charged/neutral.**

   After the transfer, chlorine has _____ positive charges and _____ negative charges.

   After the transfer, chlorine has _____ fewer positive charge(s) than negative charges.

   (Circle one answer.) So, the chlorine now has a **positive/negative** overall charge.

   Since we know that atoms are neutral in electrical charge, and we see that chlorine is no longer neutral, the resulting atom is referred to as a **chloride ion.**

(Circle one answer.) Chlorine has formed a **positive/negative** ion.

Opposite electrical charges attract. The positive sodium ions stick to the negative chlorine ions forming salt crystals. **The attraction between positively and negatively charged ions forms an ionic bond.**

ACTIVITY 3 • COVALENT BONDING

Another method of chemical bonding is called **covalent bonding.** In this type of bond, atoms **share electrons** with other atoms, thus satisfying the octet rule for all partners. This type of bond occurs when atoms need more than three electrons to complete their outermost shells or when atoms are bonding with atoms that can't form an ionic bond.

INQUIRY AND ANALYSIS

1. **Carbon** has the atomic number six. **Draw** a dot model for a carbon atom in **Figure 2-6.**

   FIGURE 2-6 Dot Model of a Carbon Atom

   Carbon needs _____ additional electron(s) to complete its outer shell.

   (Circle one answer.) Carbon will form a(an) **ionic/covalent** bond with other atoms.

2. Hydrogen has the atomic number one. **Draw** a dot model for a hydrogen atom in **Figure 2-7.**

   FIGURE 2-7 Dot Model of a Hydrogen Atom

3. Hydrogen needs _____ additional electron(s) to complete its outer shell.

   If hydrogen were to form a covalent bond with carbon, how many hydrogen atoms would be needed for all the partners to satisfy the octet rule for their outermost shells? _____

4. Return to the dot model of the carbon atom you drew in **Figure 2-6.** Add the **correct number of hydrogen atoms** to that model.

The molecule you've created has the chemical formula CH_4 (methane), which is the abbreviation for a molecule with one carbon atom and four hydrogen atoms. Since drawing dots can get tedious, covalent bonds are usually drawn with a line to represent the shared electrons. Compare the two methods of drawing the model in **Figure 2-8.**

Dot Model of Methane

Model of Methane with Lines Representing the Bonds

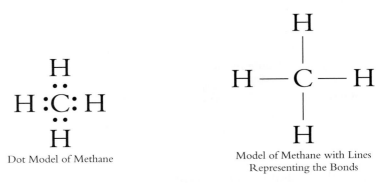

FIGURE 2-8 Comparison of Molecular Models

5. One covalent bond represents ______ electrons.

The bond between carbon and hydrogen atoms in the methane molecule, in which the atoms are sharing **one pair of electrons**, is called a **single bond.**

The methane molecule has a total of ______ single bonds.

Atoms can also share more than one pair of electrons. Sharing **two pairs** is called a **double bond.** Atoms can share a **maximum of three pairs of electrons,** resulting in a **triple bond.**

An example of multiple bonds can be found in the oxygen molecule (O_2). Oxygen has the **atomic number eight.**

6. In **Figure 2-9,** draw **two dot models** of oxygen next to each other.

Dot Model of First Oxygen Atom

Dot Model of Second Oxygen Atom

FIGURE 2-9 Dot Models of Two Oxygen Atoms

7. Oxygen needs _____ additional electrons to satisfy the octet rule.

   If two oxygen atoms form a covalent bond with each other, they would have to share _____ pairs of electrons to satisfy the octet rule for both partners.

8. Return to **Figure 2-9** and **draw a circle** that includes **one electron from each** oxygen atom.

   Draw a second circle that includes a **different electron from each** oxygen atom. Each of the circles should have **two dots** in it.

   Each circle represents a single covalent bond because it's made up of one electron pair. Since there are two of these bonds, the **oxygen molecule has a double bond.**

   The oxygen atoms are **sharing the electrons equally.** The electrons spend half their time in orbit around one partner, and the other half orbiting the other partner. This arrangement is called a **nonpolar covalent bond. Nonpolar molecules are electrically neutral.**

   If atoms **don't** share the electrons equally, this is called a **polar covalent bond.**

   In polar molecules, one atom attracts the electrons more strongly than its partners. This causes the electrons in orbit to spend more time around the stronger partner, creating a negative "pole" on that end of the molecule. The other end, in which the protons are surrounded by fewer electrons, becomes the positive pole.

    Reminder: Protons have a positive charge.

   The best known example of a polar molecule is water. In the molecular model of water in **Figure 2-10,** you can see that one end is positive and the other is negative.

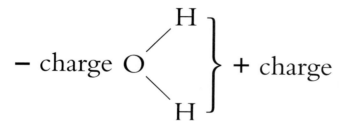

FIGURE 2-10 Molecular Model of Water

   Opposite charges attract each other. Therefore, negatively charged molecules will be attracted to, and form bonds with, the "positive" pole of the water molecule and vice versa.

   So, **any molecule that's charged will dissolve in water** (will be **water soluble**). Molecules **without a charge won't dissolve in water** (will **not** be water soluble). Perform the following experiment to illustrate this principle.

9. To **1/2 cup of water,** add **two tablespoons** of **vegetable oil.**

   To another **1/2 cup of water,** add **one teaspoon** of **food coloring.** (**Note:** If you don't have food coloring, **soy sauce** will also work.)

Based on your observations, you could conclude that water has _____ covalent bonds and oil has _____ covalent bonds.

On the other hand, food coloring (or soy sauce) probably has _____ covalent bonds.

Explain your answer.

ACTIVITY 4 • ACIDS, BASES, AND pH

When added to water, ionic molecules split apart into positive and negative ions. If the positive ion is a hydrogen ion, that substance is called an **acid.** In other words, acids are substances that release **hydrogen ions** (H^+) to a solution. The greater the concentration of H^+ released, the stronger the acid. **Bases,** on the other hand, absorb H^+ from solutions. The more H^+ they are able to absorb, the stronger the base.

The strength of acids and bases is measured by the **pH scale,** which runs from 0–14. Substances above seven on the pH scale are bases and below seven, acids. Human body fluids, for example, have a slightly basic pH (7.4). You can determine the pH of a liquid with pH test paper.

INQUIRY AND ANALYSIS

1. Test the following four household substances with the pH paper provided in your laboratory kit.

   Dip the pH paper in the liquid to be tested. Match the color the pH paper turns with the **color-coded scale** provided by your instructor.

TABLE 2-1 pH Testing

Substance Tested	pH Results (pH number)	Acid / Base / Neutral
cola		
vinegar		
milk		
liquid detergent or dish soap		

2. Test four additional substances (your choice) and record the results of your tests in **Table 2-1**.

3. Pour **1/4 cup of vinegar** into a small bowl. Place **one antacid tablet** in the bowl. Let the tablet completely dissolve (this may take up to 10 minutes) and test the solution with **pH paper**.

Record your results below:

pH of vinegar and antacid_____

pH of vinegar alone (as recorded in **Table 2-1**)_____

4. Using your knowledge of pH, which of the solutions contains **the fewest** hydrogen ions? **Explain your answer.**

5. Based on your knowledge of pH, would you classify the tablet as an **acid or a base**? **Explain your answer.**

ACTIVITY 5 • MONOMERS AND POLYMERS

Most organic compounds are made of small molecules (**monomers**) joined together into larger structures called **polymers.** Monomers are joined together by a process called **dehydration synthesis.** Synthesis refers to the joining of small components into a larger molecule. Dehydration refers to the fact that water is given off as a waste product of the synthesis process. **One water molecule is released for each covalent bond** that's formed. Most of the organic molecules that make up your tissues and organs, and that are used in body chemistry, are polymers (e.g., fats and proteins). Where do all these monomers come from? For the most part, they come from the food you eat. The digestive system breaks the polymers apart into their component monomers using a process called **hydrolysis.** Hydrolysis requires the addition of water. **One water molecule is absorbed for each covalent bond** that's broken.

The following activity will demonstrate the processes used in dehydration synthesis and hydrolysis.

INQUIRY AND ANALYSIS

1. Take **seven** paper clips and hook them together into a chain.

2. Answer the following questions using these vocabulary words: **monomer, polymer, hydrolysis, dehydration synthesis,** and **water.**

   Each paper clip represents a _____ .

   The chain of paper clips represents a _____.

   If you remove **three** paper clips from your chain, you're demonstrating the process of _____.

   One molecule of _____ will be released every time you add a paper clip to your chain.

   Adding paper clips to your chain demonstrates the process of _____.

3. How many water molecules were released when you made your chain of seven paper clips? _____

   How many water molecules would be needed in order to break your chain apart into individual paper clips? _____

ACTIVITY 6 • APPLICATIONS OF HYDROLYSIS

The sugar lactose is a disaccharide found in dairy products (such as milk). **Lactose intolerance** is the inability to complete the hydrolysis of lactose into its two monosaccharides: glucose and galactose. This occurs when a person does not produce enough of the **enzyme lactase,** which is necessary for the hydrolysis of lactose. Lactose intolerance is common in African Americans, Asians, and Hispanics. It's also possible to develop this condition as you age.

INQUIRY AND ANALYSIS

1. Pour **1/4 cup of cow's milk** into a small bowl. Dip a **glucose test strip** from your lab kit into the milk and let the test strip **develop for two minutes.** At the end of the two-minute period, **record** your results below.

   Note: Since you can't see the molecules with your naked eye, we'll be using a specialized type of chemical indicator. A glucose test strip changes from **green to brown** in the presence of **glucose.**

   Glucose test results (+ or –) _____

Since you know that lactose is composed of two monosaccharides (glucose and galactose), shouldn't your test results be positive? If your results **weren't** positive, why not?

2. Crush the **Lactaid® tablet** (a lactase enzyme supplement) that came in your lab kit and **add it to the bowl of milk.** Stir until the pieces of tablet dissolve in the milk.

   Wait **five minutes**, then repeat the glucose test using a **fresh test strip.** Let the test strip **develop for two minutes.** At the end of the two-minute period, **record** your results below.

   Glucose test results (+ or –) _____

3. Did the addition of Lactaid® change your results? If so, why?

4. **(Circle one answer.)** In this experiment, which was the control: **milk without Lactaid®/milk with Lactaid®**?

5. What would you expect to happen if you repeated this experiment with soy milk? **Explain your answer.**

24   EXERCISE 2 • INTRODUCTION TO CHEMISTRY

Exercise 3 • Organic Molecules and Enzyme Action

OBJECTIVES

After completing this exercise, you should be able to:

- test foods for the presence of glucose, starch, protein, and lipids using indicators and test strips
- relate the organic content of foods to health and dietary choices
- explain the process of hydrolysis as it relates to digestion
- describe the role of enzymes in the digestive system
- list factors that could cause proteins to be denatured

SUPPLIES

Activity 1

SUPPLIES FROM LAB KIT

- glucose test strips, 7 strips
- protein test strips, 7 strips
- pipettes, 1

HOUSEHOLD SUPPLIES

tuna packed in water, 1 TBS

whole milk, 2 TBS

potato chips, 3

smooth peanut butter, 1 TBS

raw egg, 1

ketchup, 1 TBS

tap water, 1 cup

povidone iodine 10% solution (**not** tincture of iodine), 2 oz

small bowls, 7

brown paper bag, 1

Activity 2

SUPPLIES FROM LAB KIT

- none needed

HOUSEHOLD SUPPLIES

gelatin (Jello™ or other brand), any flavor, 2 boxes

fresh pineapple, 1 piece

two other types of fresh fruit (your choice), 1 piece each

clear glass or plastic containers, at least three inches tall (for example, a juice glass), 4

tap water, 6 cups

mixing bowl, large, 1

measuring cups, 1 set

metric ruler, 30 cm (12 inch), 1

ACTIVITY 1 • ORGANIC MOLECULES IN THE FOODS YOU EAT

Any food you select probably contains several types of organic molecules, although the amount in each type of food will vary. Knowing which foods would be a good source of protein, or contain large amounts of fat, for example, can help you make healthy diet choices. In this activity, you'll be testing common foods for the presence of two types of carbohydrates (sugar and starch), protein, and fat.

 Note: In this experiment, you'll be using three specialized types of chemical indicators:

- **Iodine** changes from **reddish-brown to black** in the presence of **starch.**
- A **glucose test strip** changes from **green to brown** in the presence of **glucose.**
- A **protein test strip** changes color on a scale from **yellow to green,** depending on the protein concentration in the solution.

INQUIRY AND ANALYSIS

1. Assemble seven small bowls. Prepare each food for testing according to the directions below, placing each food into a separate clean bowl:

    - tuna: open the can, smash one tablespoon of tuna into a paste and dilute with two tablespoons of tap water.
    - potato chip: smash the potato chip into crumbs and mix it with two tablespoons of tap water.

- peanut butter: blend one tablespoon of peanut butter with two tablespoons of tap water.
- ketchup: mix one tablespoon of ketchup with two tablespoons of tap water.
- milk: no additional preparation is needed.
- egg: carefully separate the egg yolk from the white and place them **into different bowls.** No additional preparation is needed.

2. Using **glucose test strips** from your lab kit, test each of the seven foods for glucose by dipping a new strip into each bowl. **Record** your glucose test results in **Table 3-1**.

3. Using the **protein test strips** from your lab kit, test each of the seven foods for protein by dipping a new strip into each bowl. **Record** your protein test results in **Table 3-1**.

 Note: Match the color of the protein strip after the test with the **color-coded scale** provided by your instructor.

TABLE 3-1 Results of Food Analysis

Food	Glucose Strip (+ / –)	Protein Strip (+ / –)	Spot Test for Fat (+ / –)	Iodine Test for Starch (+ / –)
tuna				
potato chip				
peanut butter				
ketchup				
milk				
egg yolk				
egg white				

4. Complete the following steps to perform the spot test for lipids on each of the foods.
   - Cut one side of a brown paper bag into seven squares. **Label each square** with the name of the food to be tested.
   - Place a few drops of the food to be tested onto the paper and blot off the excess liquid. Let the spot air-dry.

 Note: For the tuna and potato chip, make sure to rub the food (not just the liquid) into the paper.

   - After each food has dried, hold the piece of paper up to the light and look for a "grease" spot. The grease spot will only be visible after the water evaporates. **Record** your spot test results in **Table 3-1.**

5. Fill the disposable pipette from your lab kit with iodine and add that amount to each food sample. **Record** your results in **Table 3-1.**

6. List **all** of the tested foods that were positive for **carbohydrates.**

7. Based on your test results, why is a peanut butter sandwich considered to be a healthy food choice for growing children? **Explain your answer.**

8. A fellow student reported an unexpected result when they tested their tuna samples. The iodine test was positive. Since he knows that starch is an energy storage molecule in plants, but not animals, he was confused. Here's the list of ingredients from the tuna can: light tuna, water, vegetable broth, and salt. **Explain** why a positive iodine test was possible in this situation.

9. If you were trying to **lower your fat intake,** which foods from the list would you eliminate from your diet? **Explain your answer.**

ACTIVITY 2 • CHEMICAL DIGESTION BY ENZYMES

For the body to make use of the foods we eat, they must first be broken down into small molecules that can be carried in the bloodstream. During digestion, large molecules are broken into their component monomers. The breakdown process is called hydrolysis, and it's accomplished through the use of digestive enzymes.

In the following experiment, you'll determine the effect of enzymes on the protein structure of gelatin. Gelatin contains the structural protein **collagen,** which causes it to solidify. Collagen is also a structural component of human skin, hair, and nails.

Proteins have a specific three-dimensional structure that allows them to perform their functions. An irreversible disruption of the three-dimensional structure of a protein is referred to as **denaturation.** If a protein is denatured, this can also change the appearance and action of the protein.

Proteins can be denatured through exposure to heat, acids, enzymes, and other environmental factors.

INQUIRY AND ANALYSIS

1. Using a piece of masking tape, number four clear containers 1–4.

2. Prepare the gelatin by following the directions on the package. SAVE THE BOX!

   Pour equal amounts of the liquid gelatin into each of the four numbered containers, as shown in **Figure 3-1.** Place the gelatin into the refrigerator to set.

FIGURE 3-1 Gelatin Setup

3. After the gelatin is firmly set, prepare the fruit.

    Note: Leave the gelatin **in the refrigerator** while you're preparing the fruit.

   Peel the pineapple and cut a piece approximately the size of a quarter. Peel the other two fruits and cut a quarter-sized piece of each.

Why is it important to make sure that the pieces of all three fruits are the **same size and shape**?

 Hint: What are the principles of valid experimental design?

4. Remove all the gelatin containers from the refrigerator.

   Place the piece of pineapple on the surface of the gelatin in glass number one.

   Place a piece of each of the other fruits into glasses two and three.

   Record the types of fruit you used in **Table 3–2.**

TABLE 3-2	Results of Gelatin Experiment			
Expired time (minutes)	Pineapple			No fruit
0:15				
0:30				
0:45				
1:00				
1:15				
1:30				
1:45				
2:00				

5. Leave the glasses of gelatin on the counter. After **15 minutes,** examine the four gelatin containers and observe the appearance of the gelatin.

   Is the gelatin in each of the four containers solid all the way from top to bottom? _____

   If not, **describe your observations on a sheet of paper,** listing each container separately.

6. If the fruit has penetrated the surface of the gelatin, measure (in **centimeters**) the distance the fruit has moved. **Record** your measurements in **Table 3-2.**

7. Repeat the procedures in step 6 (above) **every 15 minutes for 2 hours. Record** the results of each "fruit depth" observation period in **Table 3-2.**

Also, on the sheet of paper you're using to record your observations, add your new observations of the condition of the gelatin in each container (percentage of liquid and solid).

8. Did this experiment include a **control**? If so, what was it? If not, why not?

9. Summarize the results of your experiment. What were the effects of the various fruits on the gelatin?

10. Read the preparation directions on the side of the gelatin box. Are there any fruits that shouldn't be used when preparing gelatin desserts?

11. Based on the results of your experiment, why is fresh pineapple included in that list?

12. What was in the gelatin that changed its structure (from solid to liquid) when exposed to the fresh pineapple?

13. What component of the fresh pineapple could have caused the proteins in the gelatin to denature?

14. A student repeated the same experiment with canned pineapple, and he noticed that it had no effect on the gelatin. What procedures in the canning process could have led to this result? **Explain your answer.**

15. If the gelatin was solid and became liquid after the pineapple was added, could this be related to **hydrolysis**? **Explain your answer.**

16. Albumin is a protein found in egg white. What happens to the albumin when you cook an egg? Why?

 Biotechnology Today

The distressed denim look can be seen everywhere, but the traditional method of softening denim has both environmental and commercial disadvantages. The dust, grit, and acids used in the process cause water pollution and damage expensive laundry machines.

How can the "stonewashed" look be achieved without causing environmental damage? The answer is a biological method of "stoning" that doesn't depend on pumice stones. Jeans are made of cotton, which is composed of cellulose fibers. Trichoderma, a fungus that decomposes plant material, produces an enzyme called cellulase that breaks down the cellulose molecules and weakens the cotton threads softening the look and feel of the fabric.

On an industrial scale, the naturally occurring enzyme cellulase has provided manufacturers with a safe and economically profitable method of providing consumers with softer jeans.

Exercise 4 • Cells and Cell Processes

OBJECTIVES

After completing this exercise, you should be able to:

- list and explain the similarities and differences between prokaryotic and eukaryotic cells
- list the major cellular organelles and explain the function of each
- draw and label a diagram of a eukaryotic plant cell
- distinguish among animal cells, plant cells, and protist cells

ACTIVITY 1 • CELL STRUCTURE AND FUNCTIONAL PARTS

Cells are the smallest living unit and some organisms consist of only one cell (they are **unicellular**). There are many types of unicellular organisms, but all of them are quite simple and small. Some examples, as seen through a microscope, are included in **Figure 4-1.**

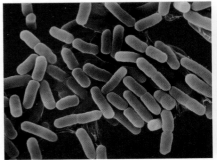
E. coli

Paramecium

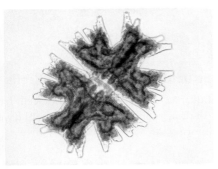
Micrasterias

FIGURE 4-1 Examples of Unicellular Organisms

Not all cells form single organisms, however. Organisms that can be easily seen with the naked eye are **multicellular** (composed of many cells). Multicellular organisms are more complex and usually consist of organized groups of cells (called **tissues**) that perform specific functions.

Within cells are specialized structures that perform specific tasks (just as various organs within your body perform specific functions). These small structures reminded the first scientists who observed them of body organs, so they were named **organelles** (which means "tiny organs"). You may be familiar with one example of an organelle—the nucleus—but cells contain many others.

Regardless of whether an organism is unicellular or multicellular, its cells will fall into either of two distinct categories: **prokaryotic** (simpler structure) or **eukaryotic** (more complicated). There are several ways to distinguish between prokaryotic and eukaryotic cells. **Table 4-1** summarizes the main differences.

TABLE 4-1	Characteristics of Prokaryotic and Eukaryotic Cells
Prokaryotic	Eukaryotic
organelles not enclosed by a membrane	many types of membrane-bound organelles
simple organization	complex organization
genetic material free-floating within cell	genetic material enclosed in a nucleus
small in size	many times larger (but still microscopic)
single-celled	often multicellular

INQUIRY AND ANALYSIS

1. **Label** each unicellular organism in **Figure 4-1** as either **prokaryotic or eukaryotic.**

2. Are your body cells prokaryotic or eukaryotic? _____ **Explain your answer.**

Structures common to all cells

Eukaryotic cells and prokaryotic cells have some organelles in common, but others are distinctive. All prokaryotic and eukaryotic cells have a cell membrane, cytoplasm, ribosomes, and genetic information.

TABLE 4-2	Structures Common to Prokaryotic and Eukaryotic Cells
Structure	Characteristics
cytoplasm	jelly-like fluid in which cell organelles are suspended
cell membrane	a) double-layered structure made of phospholipid molecules and embedded proteins b) boundary that defines the outer margin of a cell c) controls the movement of many different types of molecules in and out of the cytoplasm (nutrients, wastes, ions, etc.)
ribosomes	a) location where protein synthesis occurs b) a two-part, solid structure made of RNA and protein c) in prokaryotic cells, free floating in the cytoplasm d) in eukaryotic cells, can be free floating or attached to other organelles
genetic information	a) DNA molecule(s) b) in prokaryotes, organized into a single, coiled structure called a **chromosome** within the cytoplasm c) in eukaryotes, organized into multiple coiled chromosomes separated from the cytoplasm by the nuclear membrane

INQUIRY AND ANALYSIS

1. Using the characteristics of internal and external structures listed in **Tables 4-1 and 4-2**, **draw** a simplified prokaryotic cell and a simplified eukaryotic cell in the spaces provided below. Make your drawings large, so you can add additional parts later.

   Include and label the following structures: **cell membrane, cytoplasm, ribosomes, and genetic material.**

Prokaryotic Cell	Eukaryotic Cell

   FIGURE 4-2 Prokaryotic and Eukaryotic Cells

2. What organelle(s) do your two cells have in common?

3. In regard to your drawings, what are the most important features you used to differentiate between the prokaryotic and eukaryotic cells?

 Biotechnology Today

Hydrogen fuel cells are among the most promising new energy technologies of the future, but today, most hydrogen for fuel cells is produced using fossil fuel technology. Researchers are in the process of developing a new, green technology that uses genetically modified bacteria to create a hydrogen producing factory that's powered by sugar.

New varieties of bacteria can produce 140 times more hydrogen than would occur naturally. The hydrogen is captured and can be used to power vehicles and generate electricity. It's a clean, non-polluting fuel that releases only water and heat as waste products.

ACTIVITY 2 • ORGANELLE SCAVENGER HUNT

Eukaryotic cells are more complex than prokaryotic cells. This complexity derives from the fact that they contain more organelles and are able to perform a greater variety of functions.

INQUIRY AND ANALYSIS

1. Go to **Table 4-3**. Using information from your textbook or the Internet, **enter the name** of the cellular organelle that performs each of the functions.

2. **Figure 4-3** contains the outline of a **eukaryotic plant cell.** The outline represents the **cell membrane** and cell wall of the cell.

   Search the Internet for a photo of each organelle listed in **Table 4-3**. Using the photos you find as a guide, add a drawing of **each organelle** to the cell diagram in **Figure 4-3**.

 Hint: Develop a strategy for finding the organelles before you go running around the Internet. Some may be more difficult—it's not necessary to find the organelles in the order they're listed.

Submit your completed drawing as required by your instructor.

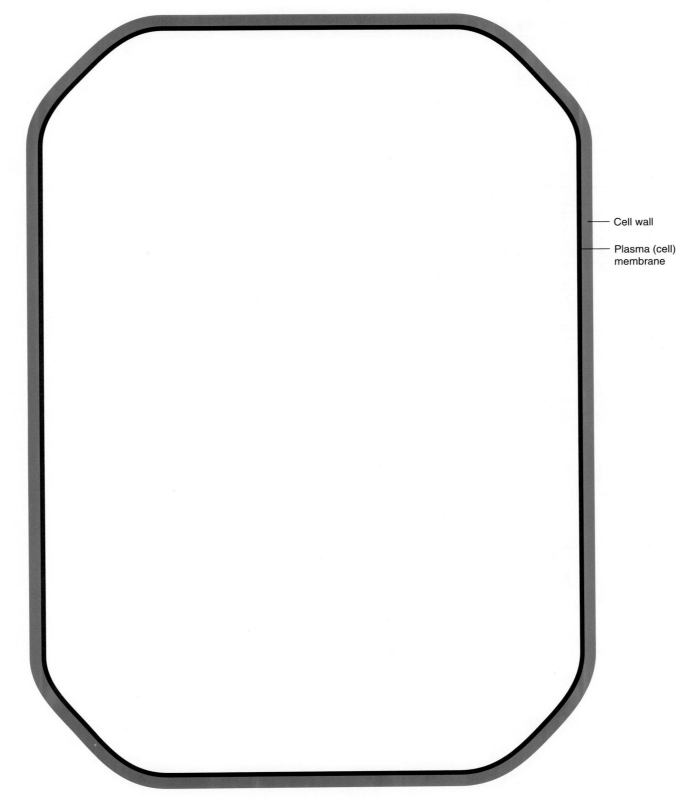

FIGURE 4-3 Diagram of a Eukaryotic Plant Cell

TABLE 4-3 Organelle Scavenger Hunt	
Organelle Function	Organelle Name
packages and exports molecules for use outside the cell in which they were made	
location where amino acids are assembled into large polymers	
location where solar energy is converted into food energy	
membrane bound compartment that encloses the chromosomes	
internal membrane system that transports materials inside the cell	
rigid exterior structure that strengthens and protects the cell	
semi-liquid "soup" that fills the space not occupied by the cell organelles	
location where food energy is converted into ATP energy	
section of the internal membrane system where carbohydrates and lipids are synthesized	
section of the internal membrane system that has ribosomes attached	
water storage organelle	

3. Which of the organelles you drew in **Figure 4-3** would **not** be found in animal cells? If there's more than one answer, **list them all.**

4. Which additional organelles, **not mentioned in Table 4-3,** are commonly found in animal cells? List at least **three** and include their **functions.**

5. If a cell blocked the entrance of a specific molecule, which organelle would perform that function?

6. You forgot to water your house plants when you went on vacation. Upon your return you notice that they're severely wilted. What cell organelle is involved?

7. Which organelle can only perform its function during the daytime? Why?

8. Bacteria also have cell walls, but they're made of different materials and have a different type of construction than plant cell walls. Several types of antibiotics target and damage the cell wall, killing the bacteria. Why are these medications safe for people to take?

ACTIVITY 3 • CLASSIFYING UNKNOWN CELLS

The junk drawer of science

You probably have an organization system in your home. Clothes go in the closet, tools in the toolbox, dishes in the kitchen, etc. But everyone has items that don't fit neatly into any particular category. Often they end up in a "junk drawer."

In a similar way, many of the unicellular organisms you can observe under the microscope are neither plant cells nor animal cells. Because they don't fit exactly into either category, they're classified into a separate group called **Protista.** They're commonly referred to as **protists.**

Some protists are more "animal-like," whereas others are more "plant-like." Some have characteristics of both plant and animal cells. For example, some cells have chloroplasts, but not a cell wall. Others that have chloroplasts can swim and detect variations in light. Some species are predators, others are scavengers. Some have fixed shapes, others can bend and flex.

To become familiar with various types of eukaryotic cells, you'll be viewing a sample of pond water through the use of a "virtual" microscope. You can find the virtual microscope on the website that accompanies this manual.

INQUIRY AND ANALYSIS

1. Go to: www.whfreeman.com/bres

   You'll find a simulation created for this activity. Follow the onscreen instructions to complete the activity.

   Click on Exercise 4 (Cells and Cell Processes).

   Click on Activity 3 (Classifying Unknown Cells).

2. View **Organism A. Write the letter A** next to each characteristic in **Table 4-4** that's present in Organism A.

   Note: You can return and look at each organism as many times as you wish.

3. Repeat step #2 for **Organisms B-H** and record your answers in **Table 4-4**.

TABLE 4-4 Unknown Cell Checklist

Diagnostic Features	Organisms with Feature Present
cell membrane present	
cell wall present	
chloroplasts present	
locomotion	
internal movement of organelles	
rigid form	
flexible form	
shape regular	
shape irregular	

4. For those cells that you said were capable of locomotion, what cellular organelle(s) **were visible** that are related to cell movement?

5. Some unicellular organisms are predators. They capture their prey by extending and enclosing the prey with extension of the cytoplasm. Which cells are capable of this process? What did you see in the microscope to support your statement?

6. Which organisms are capable of photosynthesis? **Explain your answer.**

7. Considering each organism you described as being capable of photosynthesis, which would you **not** consider to be examples of a plant cell? **Explain your reasoning.**

8. Which organisms are capable of cell respiration? **Explain your answer.**

9. Which organisms are multicellular? How did you distinguish them from the unicellular organisms?

10. If we inserted some tiny pieces of meat into the pond water culture, which of the organisms you viewed would be likely to respond? **Explain your answer.**

11. If we shaded half the pond water culture, predict the distribution of **Organism E** in the culture container. **Explain your answer.**

12. Another student in the class tells you, "It's easy to tell plant cells from animal cells. Plant cells have chloroplasts, but animal cells don't."

  a. Is this statement correct?

  b. Could you use this criterion (presence or absence of chloroplasts) to classify all the cells you saw in the pond water sample? **Explain your answer.**

Exercise 5 • Diffusion and Osmosis

OBJECTIVES

After completing this exercise, you should be able to:

- explain the concepts of diffusion and osmosis
- use chemical indicators to track molecular movement
- describe several factors that can affect the rate of diffusion
- relate the terms **hypotonic**, **isotonic**, and **hypertonic** to the process of osmosis in living cells
- explain the separation of different sized molecules through the process of dialysis

SUPPLIES

Activities 1–2

HOUSEHOLD SUPPLIES

red or blue food coloring or soy sauce, 1 tsp or a felt tip marker | water, 6 cups | container, clear, 1–2 quart volume, 1 | stop watch, (or any watch or clock with a second hand), 1

Activity 3

SUPPLIES FROM LAB KIT

- dialysis tubing, 1 piece

HOUSEHOLD SUPPLIES

water, 5 cups | salt, 1/2 cup | dental floss or string, 1 package | container, clear, 1–2 quart volume, 1

drinking glass, 1 | metric ruler, 30 cm (12 inch), 1

Activity 4

SUPPLIES FROM LAB KIT

- dialysis tubing, 1 piece
- glucose test strips, 1 strip
- pipette, 1

HOUSEHOLD SUPPLIES

honey,
1 tbsp

cornstarch,
1 tbsp

water, 8 cups

providine iodine 10%
solution (**not** tincture
of iodine), 1/2 oz.

dental floss,
1 package

container, clear
1–2 quart volume, 1

ACTIVITY 1 • DIFFUSION AT ROOM TEMPERATURE

Diffusion is a very important concept in biology. Fortunately, it's an easy concept to understand. In essence, atoms, compounds, and molecules tend to move from areas where they're highly concentrated to areas where they're not very concentrated. In other words, substances will naturally move from areas of high concentration to areas of low concentration.

You've witnessed this thousands of times in your life. Just think of what happens when someone sprays cologne or air freshener in one part of the room. If you sit still for a while, even if you're across the room, you'll smell it. That's caused by the molecules diffusing across the room from where they're very concentrated (near the spraying) into a place where they're not concentrated (far from the spraying). Diffusion happens in air, in water, and even through solids, although that takes a long time, as you'll see.

During this experiment, you'll be dropping colored liquid into a container of clear water and timing how long it takes for the color to spread through the water.

INQUIRY AND ANALYSIS

1. Take a large, clear container, and fill it with tap water. Let it sit on a table for about **15 minutes**, until **all motion in the container stops** completely.

2. From your bottle of food coloring, **gently** drip **three drops** into the water **right up against the edge of one side** of the container and **start timing immediately.**

   Note: If you don't have food coloring, you can use soy sauce or you can dunk the tip of a water-based, felt-tipped marker into the water.

   How long did it take for the water in the entire container to change from clear to colored?
   _____

   The change in water color is due to the diffusion of the food coloring.

3. **(Circle one answer.)** In your food coloring experiment:

   The container of clear water represented an area of **higher/lower** concentration of dye.

   The initial drops of food coloring represented an area of **higher/lower** concentration of dye.

4. The food coloring moved from an area of _____ **concentration** to an area of _____ **concentration.**

   The process you observed is called _____.

ACTIVITY 2 • DIFFUSION AT VARYING TEMPERATURE

In this experiment, you'll use the same procedures as in **Activity 1,** but you'll be changing the temperature of the water in the container.

INQUIRY AND ANALYSIS

1. Write a hypothesis about how the temperature of the water (cold and hot) will affect the rate of diffusion.

   Hypothesis:

2. Test your hypothesis by **repeating procedures one and two from Activity 1 twice:** first with the ice water and then with the boiling water.

   Note: For the **ice water**, remove the ice cubes from the container just before you begin your experiment.

   The important thing to remember is that you should **begin your experiment immediately after you prepare the container of water**, so that it doesn't go back to room temperature before you can complete your experiment. However, be sure to let the container sit still long enough for all water motion to stop.

   Ice water: How long did it take for the water in the entire container to change from clear to colored? _____

   Boiling water: How long did it take for the water in the entire container to change from clear to colored? _____

3. Was your hypothesis about the effect of water temperature supported? _____

4. Write a conclusion based on your hypothesis and the results of your experiment.

5. Which water temperature (room temperature, boiling water, iced water) could represent the control in your diffusion experiments? **Explain your answer.**

6. You're making iced tea. When would it be best to add the sugar: after you finish boiling the water or after you've added the ice? **Explain your answer.**

EXERCISE 5 • DIFFUSION AND OSMOSIS

7. At a restaurant, you want to sweeten your iced tea. What could you do to make the sugar diffuse through the glass faster? **Why?**

8. In addition to temperature, what other factors might affect the rate of diffusion?

ACTIVITY 3 • OSMOSIS

Now that you're an expert on diffusion, it's time to move on to osmosis. This is also an easy concept to understand. Osmosis is simply the movement of water through a membrane. That's it. It's just water moving through a **selectively permeable** membrane (the kind that surrounds all cells).

A solution has two components, the dissolved solids (called **solutes**) and the liquid (called the **solvent**). In a solution, the percentage of solute plus the percentage of solvent adds up to **100%.**

Three terms can be used to describe the relative relationship between the solute and solvent of two solutions:

- If both solutions have **exactly the same percentage of solute and solvent**, they are **isotonic** to each other.
- If one solution has a greater solute content than the other, the solution with the **higher solute percentage is hypertonic** compared to the solution with the lower solute concentration.
- The solution with the **lower solute percentage** is **hypotonic** compared to the solution with the higher solute concentration.

Using the same principles as in diffusion, the **water will always move through the membrane from an area of high concentration to an area of lower concentration.**

INQUIRY AND ANALYSIS

1. Take one of the flat pieces of dialysis tubing out of your lab kit. This is the item that's transparent, approximately six inches long, and looks like a thin strip of cellophane (see **Figure 5-1**).

FIGURE 5-1 Dialysis Tubing

2. Dialysis tubing is a synthetic **selectively permeable** membrane through which osmosis can occur (in a similar way to the cell membrane).

   Soak the tubing in a glass of room temperature water while you assemble the materials for the rest of the activity.

3. Get two pieces of **string or dental floss about six inches long.**

   Fill a **clear container** with tap water. Add **ten large spoonfuls of table salt** to the container.

   Stir the water until all of the salt is dissolved in the water.

4. Take the tubing that you've been soaking out of the water. Rub one end between your fingers until it opens (the way you open plastic bags in the produce section of the grocery store).

   You should see it open up and become a hollow tube.

   Tie one end of the tube tightly closed with a piece of string or dental floss, as shown in **Figure 5-2**, below.

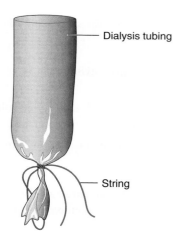

FIGURE 5-2 Dialysis Tubing Setup

5. Using one of the pipettes from the lab kit, **fill the tubing with tap water to approximately one inch from the top** of the tube.

   Tightly tie off the top, just like you tied off the bottom. Make sure there's very little air space left at the top of the tube. Your tube will look like a small sausage. Rinse the tube off with tap water.

6. **Measure the circumference of the tubing** by wrapping a piece of string around the **largest section** of the tube (see **Figure 5-3**).

   Measure the length of the string on a metric ruler and record the length: _____ cm

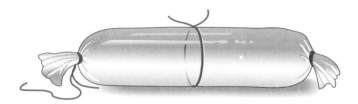

FIGURE 5-3 Measuring the Circumference of the Dialysis Tubing

7. Place the dialysis tube into the clear container of salt water. Let it sit in the water for **about four hours.**

8. At the end of the four hours, measure the circumference of the bag again, using the same procedure as in step five above.

   Measure the length of the string on a ruler and record the length: _____ cm

   Did the circumference **increase** or **decrease**? _____

9. Why did the circumference of the bag change?

   Hint: The tubing is made of a selectively permeable membrane, so you know osmosis must have happened, but you should explain the situation in terms of molecules moving in and out of the tube.

10. A salt solution that contains **20% salt** contains _____ % water.

11. If you were told that a type of cell contained about **2% salt,** what would happen if you put the cells into a solution of **18% saline**?

   (**Circle one answer.**) The cells were immersed in a **hypotonic/isotonic/hypertonic** solution.

12. What if you placed the same cells into **pure, distilled water (0% saline solution)**?

   (**Circle one answer.**) The cells were immersed in a **hypotonic/isotonic/hypertonic** solution.

13. What if you put the same cells in a **2% saline solution**?

   (**Circle one answer.**) The cells were immersed in a **hypotonic/isotonic/hypertonic** solution.

14. Based on your understanding of osmosis, why is it very important for intravenous fluids given in a hospital to have the same osmotic concentration as your own cells and body fluids?

52   EXERCISE 5 • DIFFUSION AND OSMOSIS

Biotechnology Today

The earth is 70% water, but most of that is ocean water and not drinkable. Considering the shortage of drinking water in many countries around the world, the ability to remove the salt from seawater (**desalination**) would be a great benefit to people everywhere. A biotech application of osmosis can actually accomplish this! The process is called **reverse osmosis.**

How can osmosis operate in reverse? Pressure is applied to a solution on one side of the membrane, forcing water through in the opposite direction than that of the normal movement of water in osmosis. As the process continues, pure water accumulates on one side of the membrane, leaving the salt behind.

ACTIVITY 4 • DIFFUSION AND DIALYSIS

Dialysis is the **separation of molecules** of different sizes. One way to accomplish this separation is to pass a liquid containing different-sized molecules through a piece of dialysis tubing. Molecules **smaller** than the holes (**pores**) in the tubing will **pass through.** The larger molecules will be trapped in the tubing and thus will be separated from the smaller molecules.

This experiment provides some insight into the procedures used during kidney dialysis. The kidney normally separates waste molecules from the useful molecules in the blood. In the case of kidney malfunction, a dialysis machine performs the same function.

The patient's blood flows through the dialysis machine. The machine contains a selectively permeable membrane that permits wastes and other small molecules to be removed from the blood. Blood cells and most proteins are not removed because they're too large to pass through the pores in the selectively permeable membrane.

Note: In this experiment, you'll be separating two molecules of different sizes and testing the results with two specialized types of chemical indicators.

- **Iodine** changes from **reddish-brown to black** in the presence of **starch.**
- A **glucose test strip** changes from **green to brown** in the presence of **glucose.**

INQUIRY AND ANALYSIS

1. Prepare a liquid starch solution by mixing **one tablespoon of cornstarch** with **one tablespoon of room-temperature tap water.** Mix well until the cornstarch is completely dissolved in the water.

2. Add the cornstarch and water mixture to **one cup of tap water** (again, the water should be at **room temperature**).

3. Prepare a glucose solution by mixing **one tablespoon of honey** with **one cup of room-temperature tap water.** Mix well until the honey is dissolved.

4. Take the second piece of dialysis tubing out of your lab kit. **Soak the tubing in a glass of room temperature water** while you assemble the materials for the rest of the activity.

5. Get two pieces of **string or dental floss about six inches long.**

Fill a **clear container** with tap water. Add **20 drops of iodine** to the water in the container and stir. Set this aside for later use.

6. Take the tubing that you've been soaking out of the water. Rub one end between your fingers until it opens using the same procedure you used in **Activity 3.**

Tie one end of the tube tightly closed with a piece of string or dental floss.

7. Fill the tube with approximately **one tablespoon each (10 ml)** of the **starch and glucose (honey) solutions (a total of 20 ml)**.

Tightly tie off the top of the tube, just like you tied off the bottom. Make sure there's very little air space left at the top of the tube. Rinse the tube off with tap water.

Place the filled and tied dialysis tube in the clear container with the iodine solution.

8. Record the location (inside the tube, in the container) of any changes in iodine color every **15 minutes** for an hour.

Time	Iodine Color	Location (tube or container)
Start		
15 minutes		
30 minutes		
45 minutes		
60 minutes		

9. (Circle one answer.)

Based on your observations, the starch was **able/not able** to leave the bag.

Based on your observations, the iodine was **able/not able** to enter the bag.

Explain your answers.

10. When you've completed your iodine observations, test the water in the container for glucose by dipping a **glucose test strip** from your lab kit into the water.

    Reminder: The glucose test strip changes from green to brown in the presence of glucose.

    (**Circle one answer.**) My glucose test results are **positive/negative.**

11. Based on your experimental results, which substances were able to pass through the selectively permeable membrane? How was this determined?

12. Based on your test results, which of the two molecules, starch or glucose, is larger? **Support your answer with data collected in your experiments**.

13. The tubing you used in this experiment is described as selectively permeable. What observation did you make that would lead you to believe that the tubing was, in fact, selective?

14. Explain why some substances were able to pass through the selectively permeable membrane, while others were not.

15. Dialysis is a process where substances are separated by the size of the molecules using a selectively permeable membrane. Did dialysis occur in this activity? **Support your answer with data collected in your experiments.**

Exercise 6 • Photosynthesis and Cell Respiration

OBJECTIVES

After completing this exercise, you should be able to:

- compare and contrast materials consumed and end products produced by photosynthesis and cell respiration
- use chemical indicators to test for the presence of the end products of photosynthesis and cell respiration
- predict the effect of leaf coloration on the production and distribution of starch
- discuss how environmental factors can cause changes in the rates of photosynthesis and cell respiration

SUPPLIES

Activity 1

SUPPLIES FROM LAB KIT

- pipette, 1

HOUSEHOLD SUPPLIES

- stove or hot plate, 1

freshly picked leaves from an herb or houseplant, solid green, no waxy coating (example: fresh basil), 2

rubbing alcohol (70%), ½ cup

tap water, 6 cups

povidone iodine 10% solution (**not** tincture of iodine), 2 oz

bowl, small, 1

measuring cups, 1 set

pot or saucepan, small, 1

tweezers, 1

clean, empty glass jar that can fit in the pot (such as pickle or relish jar), 1

white or light-colored plate (china or ceramic) or a jar lid (white inside and not paper lined), 1

paper towels, 1–2 sheets

EXERCISE 6 • PHOTOSYNTHESIS AND CELL RESPIRATION 57

Activity 3

SUPPLIES FROM LAB KIT

- limewater, sealed tube, 1

HOUSEHOLD SUPPLIES

drinking straw, 1

Activity 4

SUPPLIES FROM LAB KIT

- none needed

HOUSEHOLD SUPPLIES

lemon, fresh, 1

table sugar, 1 cup

tap water, 8 cups

ginger root, fresh, 5 cm (2-inch) piece

baker's yeast, 1 envelope

funnel, 1

bowl, small, 1

kitchen strainer or coffee filter, 1

two-liter plastic soft drink bottle with cap (clean), 1

vegetable grater (preferably with a set of fine-cutting teeth), 1

measuring cup (must hold at least one cup of liquid), 1

measuring spoons, 1 set

ACTIVITY 1 • THE STARCH TEST AS AN INDICATION OF PHOTOSYNTHESIS

Basically, **photosynthesis** is a process of converting solar energy into food energy. It can only be done by organisms that contain chlorophyll in their cells. This includes not only trees and other plants, but also algae and many types of bacteria.

The process of photosynthesis uses the sun's energy to produce a sugar called **glucose.** The energy in the glucose molecule can then be used to form other types of molecules, such as proteins, lipids, or complex carbohydrates that the organism needs.

For an overview of photosynthesis, see **Figure 6-1.**

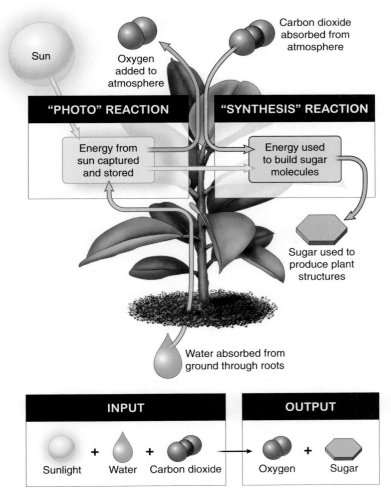

FIGURE 6-1 Overview of Photosynthesis

Since photosynthesis can only occur when sunlight is available, plants must store enough excess energy during daylight to cover their energy needs during times of reduced sunlight, darkness, or seasons of the year when leaves are absent.

The **excess glucose** formed during photosynthesis is stored for later use in the form of **starch.** Therefore, an indirect method of determining whether photosynthesis has occurred in a plant would be to test for the presence of starch. We can do this by using the indicator chemical **iodine.**

 Note: If starch is present, the iodine will **change color from reddish-brown to black**.

Four ingredients are needed for photosynthesis to take place. If you take away **any of the four,** this prevents photosynthesis from occurring.

- **Sunlight:** provides the energy used to form glucose
- **Chlorophyll:** a pigment that traps light energy
- **Carbon dioxide:** provides the carbon and oxygen atoms needed to build glucose molecules
- **Water:** provides the hydrogen atoms needed to build glucose molecules

As shown in **Figure 6-1,** a waste product of photosynthesis is **oxygen**—the same oxygen we all breathe. Without producers performing photosynthesis, life as we know it on this planet wouldn't be possible. **Half of the world's oxygen is produced by microscopic algae** in oceans, lakes, and streams. The **other 50%** is produced by trees and other plants.

Read all the instructions before you begin your experiment.

INQUIRY AND ANALYSIS

1. Pick two leaves from a healthy houseplant. Fill a pot **half way** with tap water and heat to boiling. Place the **two freshly picked leaves** into the boiling water and boil for **five minutes.**

   Note: Plant cells are surrounded by rigid support walls. The cell walls must be softened by boiling in order for the indicator to penetrate the cells.

2. While the leaves are boiling, measure **one-half cup of rubbing alcohol** and pour it into the glass jar. Set the jar aside for later use.

3. **Caution! Remove the pot from the stove** and place it on a **heat-resistant surface.** Using tweezers, remove the leaves from the pot and place them into the jar containing the rubbing alcohol. **Don't discard the hot water.**

4. **Caution!** Carefully place the jar containing the alcohol and the leaves into the pot of hot water (see **Figure 6-2**). **Make sure that no water from the pot enters the jar.**

   Don't place the pot back on the stove. Don't heat the water any further.

FIGURE 6-2 Set-up for Chlorophyll Extraction

5. Let the leaves sit in the alcohol for 15–30 minutes or until they turn **white** (see **Figure 6-3**). Observe the **color of the alcohol** in the jar. It has changed from clear to _____ . Why?

FIGURE 6-3 Appearance of Leaf Before and After Chlorophyll Extraction

6. While the leaves are in the alcohol, **prepare the rinse water** (tap water in a small bowl) and position a **light-colored plate or jar lid** near your work surface.

   Using the tweezers, carefully remove the leaves from the alcohol (taking care not to tear the leaves), and dip them in the rinse water.

   Blot the leaves gently on a paper towel or napkin and place them in different areas on the light-colored plate or jar lid. Using the tweezers, gently spread the leaves back into their former shapes.

7. Using a pipette from your lab kit, flood the leaves with iodine. After **two minutes**, observe the color of the leaves.

   Reminder: Iodine is an indicator for the presence of starch. If starch is present, the iodine will change color from **reddish-brown to black**.

   (Circle one answer.) The results of my iodine test were **positive/negative** for **starch**.

8. _____ molecules formed during the process of _____ are converted to starch for storage.

9. Based on your knowledge of photosynthesis, **list and explain at least two factors** that could lead to a **negative iodine test** in a houseplant.

10. Imagine that you tested your plant for starch with iodine and the results were positive. Then you placed the same plant in the dark for two weeks. Imagine further, that when you repeated the iodine test, the results showed that starch levels had decreased over the two-week period. **Explain** what happened to the starch. **Be specific.**

ACTIVITY 2 • THE STARCH TEST WITH VARIEGATED LEAVES

As you probably know, not all plants have solid green leaves. A leaf with different color patterns is referred to as **variegated.** In this activity, you'll be forming hypotheses about whether variegated color patterns could have an effect on the **production and distribution of starch** in leaves.

INQUIRY AND ANALYSIS

1. Look at the photos of the coleus plant in **Figure 6-4.** In this species of coleus the leaves have a green and white pattern.

(a) Coleus Plant

(b) Coleus Leaf

FIGURE 6-4 Coleus Plant

2. Imagine that you're performing an iodine test using coleus leaves. **Draw a diagram** of a coleus leaf in the box provided below. Based on your knowledge of photosynthesis from **Activity 1,** shade in those sections of the leaf where you expect to find starch.

   The shaded diagram represents **your hypothesis** about the results of an iodine test on the coleus leaf.

   Hypothesis – Shaded Diagram of a Coleus Leaf

3. Did you shade in the entire leaf? **Explain your answer.**

4. Which **section** of the leaf could serve as a **control** in this experiment? **Explain your answer.**

5. **(Circle one answer.)** The results of this experiment (as seen in **Appendix II**), **supported/didn't support** my hypothesis.

6. What necessary ingredient might be missing in areas of the leaf that exhibited a negative result to the iodine test? **Explain your answer.**

ACTIVITY 3 • CELL RESPIRATION

As we've seen, photosynthesis converts **solar energy into chemical (food) energy.** This food energy, however, must be converted one more time into a molecule that can be used by cells.

This is analogous to the conversion of computer files. Let's say you've written a report to be posted on your course web page. To create an acceptable submission, you have to convert your file to another format (e.g., from .docx to .html).

In a similar manner, your body cells can only accept energy in the form of **ATP molecules.** The energy in carbohydrates, proteins, and fats in the food you've eaten has to be converted into **ATP energy,** which serves as fuel for work in the body. The conversion process is called **cell respiration.**

During cell respiration, food energy is converted into ATP energy and the left-over atoms are combined and **released as wastes** in the form of **carbon dioxide and water** (the original ingredients used in photosynthesis).

For a summary of the cell respiration process, see **Figure 6-5.**

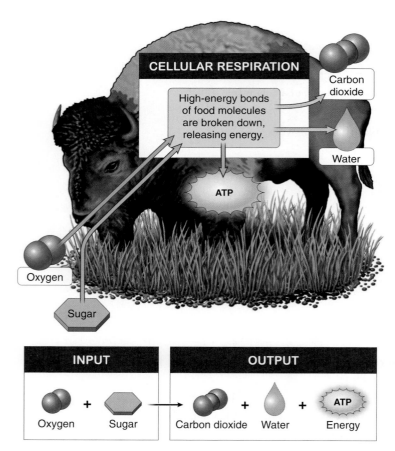

FIGURE 6-5 Overview of Cell Respiration

EXERCISE 6 • PHOTOSYNTHESIS AND CELL RESPIRATION

All living organisms on this planet, from large multicellular animals such as elephants or humans, to plants, to the smallest microorganisms use the process of cell respiration to obtain energy from their food. Try the following experiment to see the results of cell respiration.

INQUIRY AND ANALYSIS

1. Carefully remove the cap from the **tube of limewater** in your lab kit.

   Note: Limewater is an indicator chemical that changes color from **clear to cloudy white** in the presence of **carbon dioxide gas.**

2. **Very gently** bubble your exhaled air through a drinking straw into the tube of limewater, as shown in **Figure 6-6**.

   Caution! Be careful not to suck the limewater solution into your mouth.

   Continue bubbling exhaled air into the limewater for **one minute.**

FIGURE 6-6 Limewater Set-up

3. Did the limewater change color? _____

   What did the color change indicate? _____

   Was CO_2 exhaled into the limewater? _____

   If so, what biological activity produced the exhaled CO_2? _____

4. Form a hypothesis. How would the limewater test be affected if you exhaled through the straw immediately after completing a run on a treadmill? **Explain your answer.**

5. An animal on the side of the road appears to be inflated with gas. How could you determine whether the gas was a result of decomposition by microorganisms inside the carcass?

ACTIVITY 4 • CELL RESPIRATION IN MICROORGANISMS

All living organisms use the process of cell respiration to obtain energy from their food. You can demonstrate the process of **cell respiration in yeast** by making ginger ale. Yeast are **microscopic fungi** that are used for many commercial purposes.

INQUIRY AND ANALYSIS

1. Take a piece of **fresh ginger root** about two inches (5 cm) long (see **Figure 6-7**).

   Peel the brown skin off the ginger root and **finely grate.** Place the grated ginger root into a small bowl and add the **juice of one fresh lemon** (about two tablespoons of juice). Stir well to mix the ginger and the lemon juice.

FIGURE 6-7 Ginger Root

2. Using the **funnel**, pour **one cup of sugar** into a clean, two-liter bottle. Add one-quarter teaspoon of yeast to the bottle. Shake the bottle to mix the sugar and yeast.

   Carefully add the ginger root mixture to the two-liter bottle.

3. Fill the bottle with **lukewarm water. Leave three inches (7.5 cm) of air space at the top of the bottle** (see **Figure 6-8**).

   Caution! If you don't leave the air space, your bottle might burst and create a big mess.

4. **Tightly** cap the bottle. Shake the bottle until the contents are well mixed.

   Note: The ginger **won't dissolve** in the liquid.

FIGURE 6-8 Ginger Ale Set-up

5. Place the bottle on your kitchen counter in a warm location. Check the bottle by squeezing the sides with your thumb. **Check several times a day.**

   As the ginger ale forms, the carbonation will increase, and the sides of the bottle will become more and more resistant to squeezing (**24-48 hours**). If you can put a dent in the side of the bottle with your thumb, the ginger ale **isn't** ready.

   When you can no longer dent the sides of the bottle by squeezing, **put the bottle in the refrigerator. Refrigerate at least 12 hours, so the bottle is completely chilled.**

   Caution! Refrigeration is necessary to prevent the pressure of the contents from popping the cap off the bottle.

6. When the bottle is thoroughly chilled, **slowly** unscrew the cap to release the pressure. Pour the ginger ale through a kitchen strainer or coffee filter to remove the grated ginger pieces. It's ready to drink—enjoy your beverage.

7. What gas was released when you opened the bottle of ginger ale? _____

   What process produced the gas? _____

8. Based on the results of your experiment, was the yeast you added to the bottle alive? **Explain your answer.**

9. If we released the gas from the ginger ale bottle in a room full of plants, could the plants make use of the gas? **Explain your answer.**

10. What happened to the **rate** of CO_2 production when the bottle was refrigerated? **Explain your answer.**

11. In reference to your answer to question 10, how does refrigeration help prevent food from spoiling?

12. (Circle one answer.) Would the amount of carbon dioxide produced have **increased/ decreased/stayed the same** if we had used **boiling** water to make the ginger ale instead of lukewarm water? **Explain your answer.**

Exercise 7 • Mitosis and Meiosis

OBJECTIVES

After completing this exercise, you should be able to:

- correctly use terminology pertaining to cell division
- identify the stages of mitosis in onion cells as viewed through the microscope
- list and explain several factors that increase genetic diversity in a population
- discuss the effect of crossing over on inheritance
- compare and contrast mitosis and meiosis in terms of the chromosome number and genetic composition of the daughter cells produced
- construct and interpret graphs that show the effect of random assortment

ACTIVITY 1 • MITOSIS

When a cute little baby grows into a teenager, there's a dramatic increase in body size. Growth is an increase in the number of body cells. Where do all these new cells come from? When you cut your lawn, how does the grass grow back? When your skin cells flake off or your stomach lining wears out, where do the replacement cells come from?

The answers to all these questions can be found through study of a process of cell division called **mitosis.** Mitosis occurs in plants, animals, and microorganisms. It's the basic method by which new cells are produced.

When performing its normal functions in the body, a cell's physiological state is referred to as **interphase.** Interphase isn't considered to be a stage of mitosis, but it's an important part of the cell cycle. In fact, a cell spends most of its time in interphase (as shown in the diagram in **Figure 7-1**).

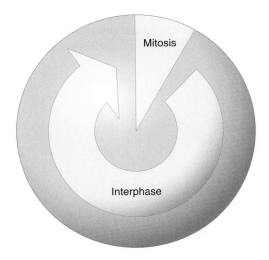

FIGURE 7-1 The Cell Cycle

The cells produced by mitosis are referred to as **daughter cells,** although the process occurs in both males and females. During interphase, a cell preparing to divide will perform several activities necessary to complete mitosis. One of the most important of these activities is the **duplication of chromosomes** (shown in **Figure 7-2**).

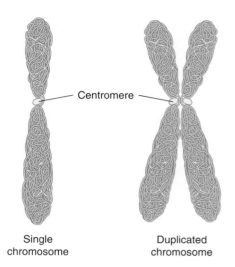

FIGURE 7-2 Duplication of Chromosomes

It's critical that new cells contain all the **genetic information** required to become fully functioning body cells. To complete mitosis, therefore, **two complete sets of chromosomes** are required: one set for each of the daughter cells formed. The genetic information in each cell is divided into **chromosomes.** The particular number of chromosomes for each species is called the **diploid** number (abbreviated **2n**).

During the stages of cell division, **the two sets of chromosomes are separated and distributed** to the two daughter cells. Since the chromosomes are **duplicate sets,** the daughter cells will also be identical duplicates of the original parent cell. **A diploid parent cell results in two diploid daughter cells** (see **Figure 7-3**). Thus, all the cells of your body are genetically identical to each other and work together as a cohesive whole.

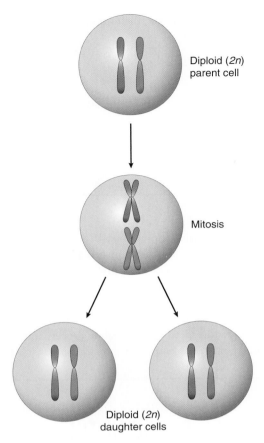

FIGURE 7-3 Production of Diploid Daughter Cells

INQUIRY AND ANALYSIS

1. If an organism has a **diploid chromosome number of 20** and goes through **mitosis,** how many chromosomes will be present in each of the daughter cells? _____

2. Why does the rate of chromosome duplication increase when you get a cut?

3. What would happen if mitosis could no longer occur in an **adult** plant or animal? Would this be a problem? **Explain your answer.**

4. Predict the relative rate of mitosis in a tree in your backyard during the four seasons of the year. **Explain your answer.**

Biotechnology Today

People with extensive burns are in a difficult situation, especially if more than 50% of their skin has been damaged. They need millions of new skin cells to repair the burned areas. A new technique involves taking a small piece of healthy skin from a burn victim and using it to grow new cells in a laboratory. In a short time, a lab can grow enough cells (by the process of mitosis) to cover the burned area.

When the lab grown cells are sprayed on a burn victim's damaged skin, they continue to divide. Preliminary results show that this new technique can speed up the healing process and reduce scarring. Although this treatment is still undergoing clinical trials, it shows great promise for the future.

ACTIVITY 2 • STAGES OF THE CELL CYCLE IN THE ONION ROOT TIP

A green onion plant has three main sections: the leaves, the stem, and the roots. Just as the leaves continue to grow (and perform photosynthesis), the roots also grow and spread, seeking out minerals and water in the soil.

A green onion plant is shown in **Figure 7-4b.**

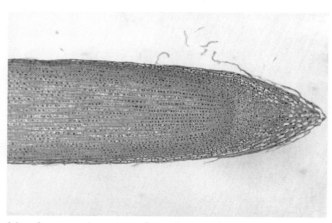

(a) onion root tip as seen through the microscope **(b)** parts of the onion plant

FIGURE 7-4 Green Onion Plant

New cells for growth are produced by mitosis. If you cut a very thin piece of the root tip and stain it (as shown in **Figure 7-4a**), you can actually see the cells going through the stages of mitosis. Amazing!

The process of mitosis is divided into a series of continuous stages. Each stage blends into the following stage, but it is possible to characterize each stage by specific events. The stages occur in sequence. **Table 7-1** shows the events that distinguish each of the four stages of mitosis.

TABLE 7-1   Distinguishing Events of Mitosis

Stage of Mitosis	Events
prophase	chromosomes become darker and more visible
	the nuclear membrane becomes less visible
	mitotic spindle forms and chromosomes attach to spindle fibers
metaphase	chromosomes line up in the center of the cell
anaphase	chromosomes begin to separate into two distinct groups, pulled apart by the spindle fibers
telophase	the separation of chromosomes is complete
	cytokinesis (splitting of cytoplasm) into two new cells
	a new cell wall forms (plant cells only)

Figure 7-5 shows the appearance of onion cells during each of the four stages of mitosis and also during interphase.

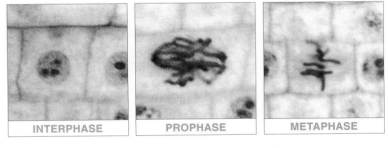
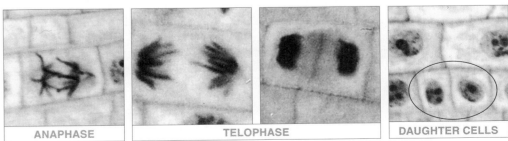

FIGURE 7-5 Interphase and the Stages of Mitosis in Onion Cells

INQUIRY AND ANALYSIS

1. Based on the descriptions of the stages of mitosis in **Table 7-1**, label the following events in the photographs in **Figure 7-5**:

   - new cell wall starts to form
   - first appearance of individual chromosomes
   - start of separation of two sets of chromosomes
   - chromosomes line up in the center of the cell
   - daughter cells formed
   - cell not undergoing mitosis

2. In each of the eight photographs in **Figures 7-6** through **7-9**, count the number of cells in interphase, prophase, metaphase, anaphase, and telophase.

   In these photographs of the onion root tip, taken through the microscope, **each box is a separate cell.**

   Count **only** those cells in which a **nucleus or chromosomes are visible.** Record your results in **Table 7-2.**

TABLE 7-2 Counting Cells in the Onion Root Tip

	Number of Cells in Each Stage of the Cell Cycle				
	Interphase	Prophase	Metaphase	Anaphase	Telophase
Figure 7-6a					
Figure 7-6b					
Figure 7-7a					
Figure 7-7b					
Figure 7-8a					
Figure 7-8b					
Figure 7-9a					
Figure 7-9b					
Totals					

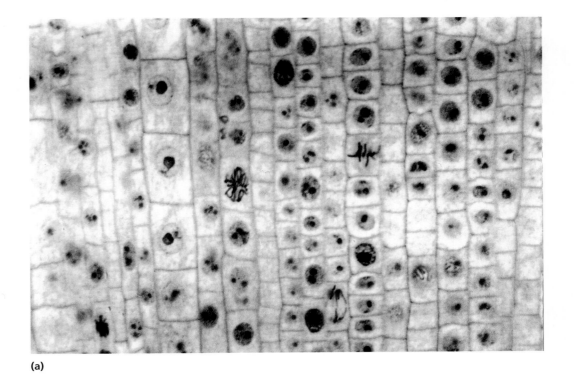

(a)

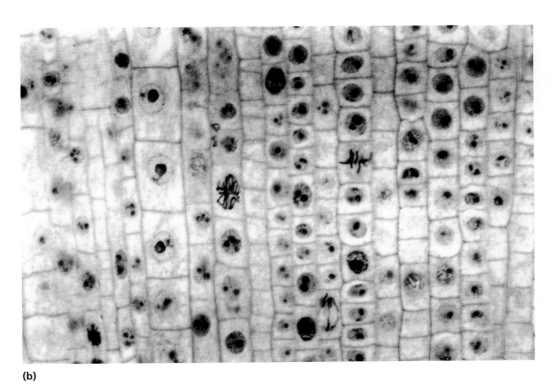

(b)

FIGURE 7-6 Cells of the Onion Root Tip (a) *top*, and (b) *bottom*, Part I

ACTIVITY 2 • STAGES OF THE CELL CYCLE IN THE ONION ROOT TIP **75**

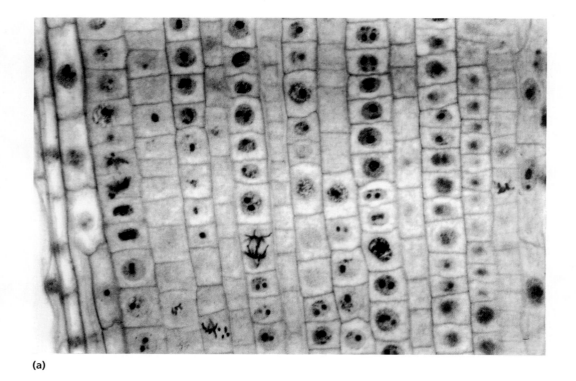
(a)

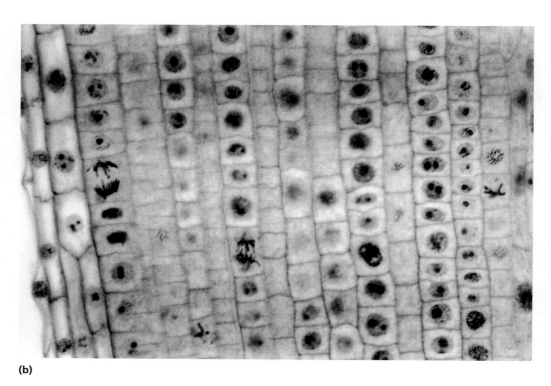
(b)

FIGURE 7-7 Cells of the Onion Root Tip (a) *top*, and (b) *bottom*, Part II

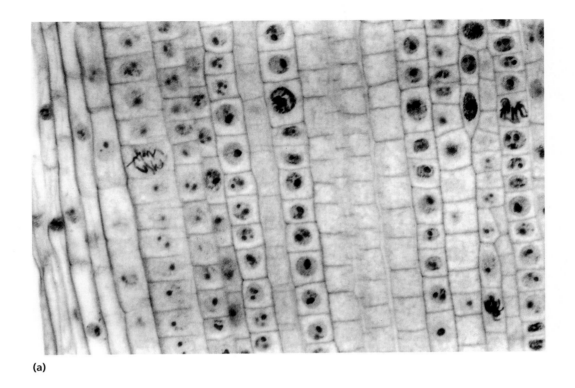

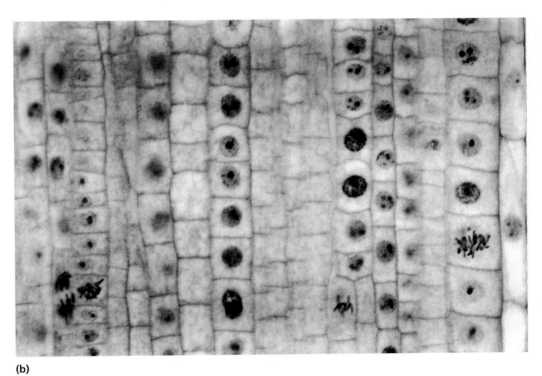

FIGURE 7-8 Cells of the Onion Root Tip (a) *top*, and (b) *bottom*, Part III

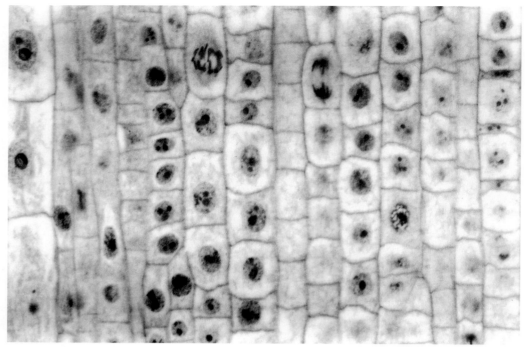

(a)

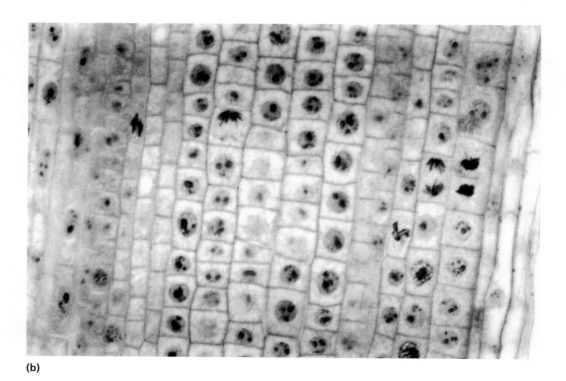

(b)

FIGURE 7-9 Cells of the Onion Root Tip (a) *top,* and (b) *bottom,* Part IV

3. Which stage of the **cell cycle** is the **mode**? _____

   ✋ **Reminder:** The mode is the stage of the cell cycle that occurs most frequently.

4. Which stage of **mitosis** is the mode? _____

5. Based on the number of cells you counted in each stage of the **cell cycle,** in which stage does a cell spend most of its time? _____

6. Based on the number of cells you counted in each stage of **mitosis,** which stage takes the longest to complete? _____

7. If you were looking at the photograph in **Figure 7-8a three weeks after** your initial count, would the distribution of the stages appear exactly the same? **Explain your answer.**

8. If you examined the root tip of an oak tree in the fall, would you expect to see the same rate of mitosis as in the spring? **Explain your answer.**

ACTIVITY 3 • INTRODUCTION TO MEIOSIS

New **body cells** are produced by **mitosis.** However, there's another type of cell division that's used to produce sex cells (sperm and eggs). As opposed to mitosis, this second type of cell division, called **meiosis,** produces cells that are genetically quite different from each other.

You only have to look around to notice that there's a lot of diversity among humans, animals, and even plants. Organisms differ from others of the same species not only in their physical appearance, but also in their physiological characteristics (such as blood type and enzyme production). This diversity arises from underlying genetic differences between individuals.

In terms of the survival of a species, genetic diversity is beneficial. A diverse population is better able to cope with and survive changes in the environment, spread of diseases, and other unpredictable events that may occur over time. Even simple organisms such as bacteria have a method of shuffling genes. Thus they evolve and change over generations. An example is the recent increase in types of antibiotic resistant bacteria.

During meiosis, **half of an individual's genetic information comes from each parent.** This results in offspring that are **not identical to their parents.** Therefore, meiosis acts as a mechanism to ensure and increase genetic diversity in a species.

Within a cell, chromosomes come in pairs. One member of each pair is inherited from your mother (the **maternal chromosome**) and the other from your father (the **paternal chromosome**). A human cell, which has **46** chromosomes, forms **23 pairs.** Each pair has one maternal and one paternal chromosome (see **Figure 7-10**).

Sperm and eggs, which are produced by **meiosis,** contain **only half the chromosomes** of a normal body cell. That's why you get only half of your chromosomes from each parent. Since the **normal** number of chromosomes is referred to as the **diploid (2n)** number, cells with only **half the normal number** are called **haploid. Cells produced by meiosis are always haploid (n).**

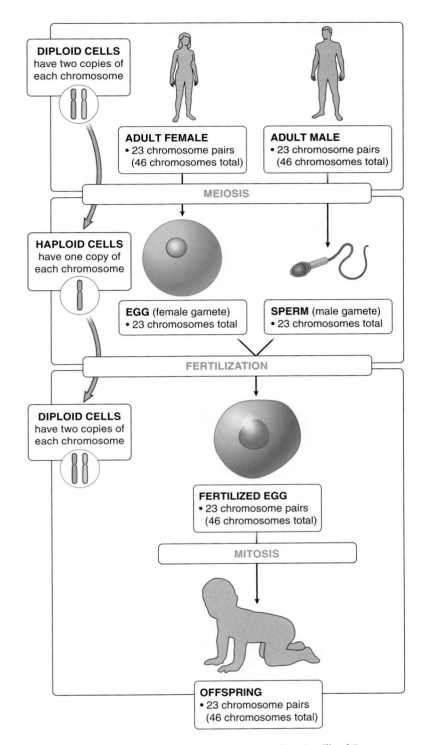

FIGURE 7-10 Inheritance of Chromosomes in a Fertilized Egg

As you saw in the previous activities, a cell preparing to undergo mitosis **duplicates its chromosomes** during interphase. This **also occurs** in preparation for **meiosis.** As a result of this duplication, each maternal and paternal chromosome produces an extra copy. The duplicates in both mitosis and meiosis are known as **sister chromatids.**

Figure 7-11 has an example of this duplication using a "model" cell with only four chromosomes.

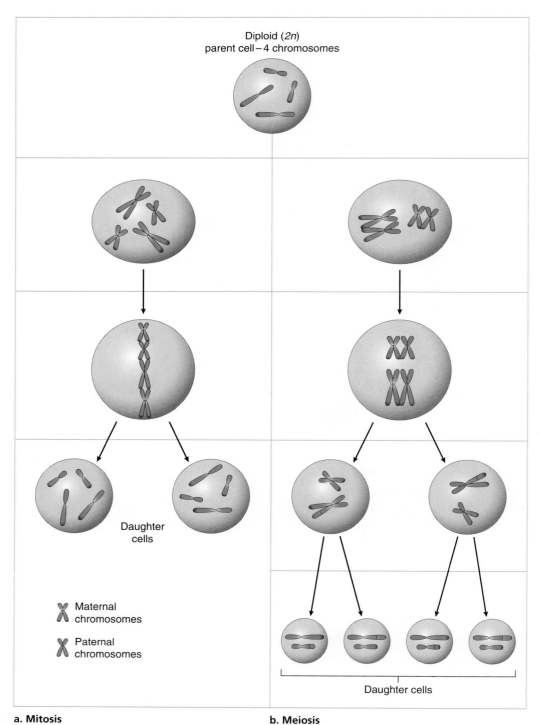

FIGURE 7-11 Comparison of Daughter Cells Produced by Mitosis and Meiosis

INQUIRY AND ANALYSIS

1. In a cell with four chromosomes, there would be _____ pairs of chromosomes.

   Each of these pairs would have _____ maternal and _____ paternal chromosome(s).

2. In **Figure 7-11a** (mitosis), the parent cell contains _____ chromosomes and each daughter cell contains _____ chromosomes.

   (Circle one answer.)

   The daughter cells produced by **mitosis** are genetically **identical to / different from** each other.

   The daughter cells produced by **mitosis** are genetically **identical to / different from** the parent cell.

   The daughter cells produced by **mitosis** are **diploid / haploid**.

   The daughter cells produced by **meiosis** are **diploid / haploid**.

3. In **Figure 7-11b** (meiosis), the parent cell contains _____ chromosomes and each daughter cell contains _____ chromosomes.

   (Circle one answer.)

   The daughter cells produced by **meiosis** are genetically **identical to / different from** each other.

   The daughter cells produced by **meiosis** are genetically **identical to / different from** the parent cell.

   Each of the daughter cells contains **one / both** of the chromosomes from each pair.

4. If the diploid parent cell in **Figure 7-11b** had **eight chromosomes,** how many chromosomes would be present in each daughter cell? _____

5. Because of a fertility problem, your eggs were fertilized in vitro and several embryos were frozen. If you examined one of the frozen embryos, would you expect **mitosis or meiosis** to be occurring in the cells? **Explain your answer.**

ACTIVITY 4 • GENETIC DIVERSITY

Although meiosis occurs in the same four stages as mitosis (prophase, metaphase, anaphase, and telophase), in meiosis the sequence of stages is repeated **twice** (see **Figure 7-12**). For this reason, the stages are often represented by Roman numerals (such as prophase I and prophase II).

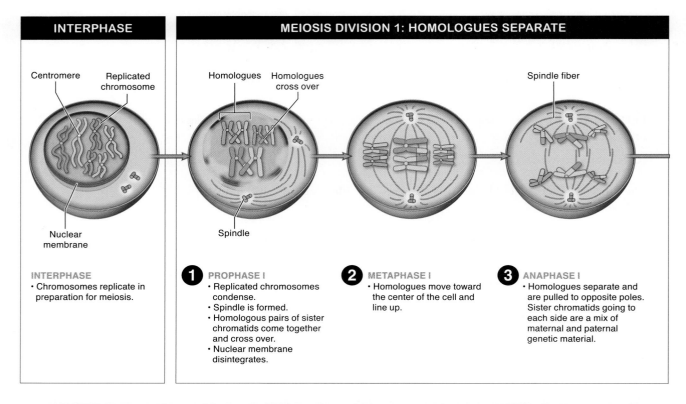
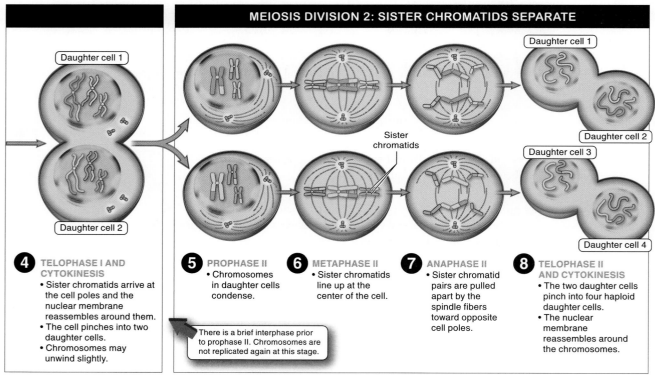

FIGURE 7-12 Interphase and the Divisions of Meiosis

During meiosis, genes are shuffled and mixed during the production of sperm and eggs. The process is called **crossing over.** This adds to genetic diversity.

Crossing over takes place during **prophase I.** During crossing over, there is an exchange of genetic information **between the maternal and paternal chromosomes** within each pair (see **Figure 7-13**).

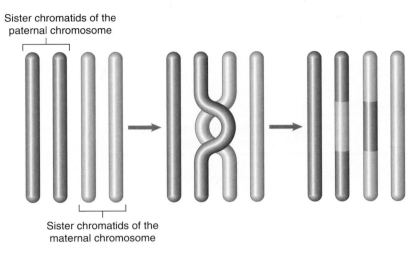

FIGURE 7-13 Crossing Over

INQUIRY AND ANALYSIS

1. **Before** crossing over occurs, are the **two chromatids of the maternal chromosome** identical? _____

   After crossing over occurs, are the **two chromatids of the maternal chromosome** identical? _____

   Explain your answers.

2. **(Circle one answer.)** The event that ensures that a baby will have the normal **(diploid)** number of chromosomes is **ovulation / fertilization / implantation. Explain your answer.**

If you look carefully at **Figure 7-11b,** you'll notice that, although each daughter cell ends up with two chromosomes, they all received different combinations of maternal and paternal chromatids. In fact, there's no way to predict which random combination of maternal and paternal chromosomes will be found in a specific sperm or egg. The random distribution of chromosomes to the daughter cells (called **random assortment**) acts to **increase genetic diversity.**

The example in **Figure 7-11b** only showed two pairs of chromosomes, but humans have many more. Each cell has **23 pairs** of chromosomes, all of which are randomly assorted in the various sperm and eggs. This dramatically increases the possible mixing of genes and, therefore, the genetic diversity of the next generation.

Table 7-3 lists the number of genetically different gametes that can be produced as more chromosome pairs are randomly assorted.

TABLE 7-3 Effect of Random Assortment

Number of Chromosome Pairs	Possible Number of Genetically Different Gametes
1	2
2	4
3	8
4	16
5	32
6	64
7	128
8	256
9	512
10	1,024
11	2,048
12	4,096
13	8,192
14	16,384
15	32,768
16	65,536
17	131,072
18	262,144
19	524,288
20	1,048,576
21	2,097,152
22	4,194,304
23	8,388,608

3. **Create a graph** showing the information in the table. **Prepare the graph on a computer** using the instructions in **Appendix I.**

   Submit the graph as required by your instructor.

4. **(Circle one answer.)** As the number of chromosome pairs increases, the number of possible gametes **increases / decreases / stays the same.**

5. Based **only on the effects of random assortment,** how many possible different genetic combinations exist each time an egg is fertilized? _____

6. John and Alice Smith have five sons that are quite different in appearance and talents. Use the concept of **random assortment** to help explain why this occurs. **Explain in detail.**

7. **List and explain** the two events that occur in meiosis that act to increase genetic diversity within a species.

   a.

   b.

Exercise 8 • Human Genetics

OBJECTIVES

After completing this exercise, you should be able to:

- interpret male and female karyotypes
- explain possible causes of common chromosomal abnormalities
- define, explain, and properly use genetic terms in context
- determine the genotypes and phenotypes of family members and use them to solve genetics problems
- use a Punnett Square to determine the probability of transmission of specific traits to offspring
- determine the most probable mode of inheritance for a trait based on experimental data

ACTIVITY 1 • UNDERSTANDING GENETICS THROUGH CHROMOSOME ANALYSIS

Humans have 46 chromosomes in their body cells (with the exception of mature red blood cells). This is referred to as the **diploid (2n)** number of chromosomes. The diploid number of chromosomes actually consists of **two pairs of 23 chromosomes.** One set of 23 was inherited from your father (the **paternal** chromosomes) and the other 23 from your mother (**maternal** chromosomes). Each maternal chromosome has a matching paternal partner. The two chromosomes that form a pair are called **homologous chromosomes.** So, in a human cell, there are 23 homologous pairs of chromosomes.

A cell preparing to carry out meiosis **duplicates its chromosomes** during interphase. As a result of this duplication, each maternal and paternal chromosome produces an extra copy. The duplicates are known as **sister chromatids.**

You can see chromosomes through a light microscope, but they're only visible during the process of cell division. It's possible to "freeze" cell division at a particular stage so that the chromosomes can be photographed and counted. The scanned images can be manipulated by computer into an orderly set showing each pair of homologous chromosomes. This arrangement is called a **karyotype.**

The chromosome pairs are assigned numbers (1–22) and are referred to as **autosomes.** The last pair is referred to as the **sex chromosomes.** These are designated with letters of the alphabet, X and Y. A female has two X chromosomes, and is designated XX. A male has one X and one Y chromosome, and is designated XY.

INQUIRY AND ANALYSIS

1. **Figure 8-1** shows an example of a **female** karyotype.

    Are there 46 total chromosomes? _____

    How many pairs of **autosomes** are in the karyotype? _____

    How many pairs of **sex chromosomes**? _____

2. Examine the homologous pair of sex chromosomes (**Figure 8-1**). Is there a visible difference between the maternal and paternal chromosomes? _____ **Explain your answer.**

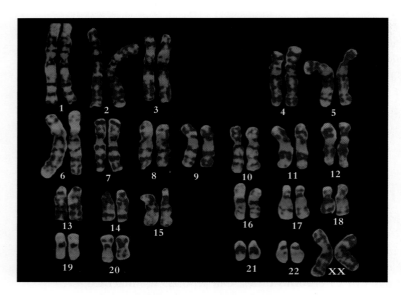

FIGURE 8-1 Female Karyotype

3. **Figure 8-2** shows an example of a **male** karyotype.

Are there 46 total chromosomes? _____

How many pairs of **autosomes** are in the karyotype? _____

How many pairs of **sex chromosomes**? _____

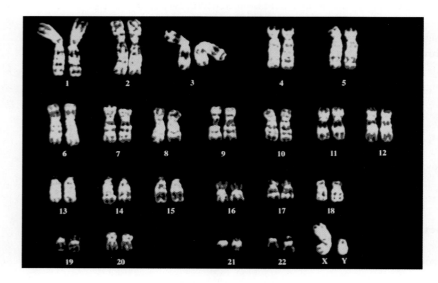

FIGURE 8-2 Male Karyotype

4. Is there a visible difference between the **X and Y** chromosomes? _____ **Explain your answer.**

 Biotechnology Today

Where did the human race originate? Which pathways did your ancestors follow when they migrated to other continents? The Genographic Project is a joint venture between the National Geographic Society and IBM that's finding the answers to these questions. Since the project began in 2005, DNA has been contributed by hundreds of thousands of people around the world. This DNA has been analyzed by geneticists and computational biologists (a new career opportunity!). What has the project discovered so far? Regardless of where in the world the DNA samples were collected, all our genetic roots can be traced back to East Africa.

To learn more about the genetic history of the human race and for information on how to contribute your own DNA to the project, visit https://genographic.nationalgeographic.com/genographic/index.html

ACTIVITY 2 • ABNORMAL KARYOTYPES

One purpose of constructing karyotypes is that it allows us to screen for genetic disorders that are caused by an abnormal number of chromosomes in the fertilized egg. One type of error that can occur during gamete formation is called **nondisjunction.** Human sperm and eggs normally have the **haploid** number of 23 chromsomes (n). Nondisjunction can result in a sperm or egg with too many or too few chromosomes.

Nondisjunction can occur **in either meiosis I or meiosis II** (see **Figures 8-3 and 8-4**). An abnormal chromosome number occurs when homologous chromosomes or sister chromatids don't separate in the production of sperm or eggs. If an abnormal sperm or egg is involved in fertilization, the resulting zygote would not form with the proper diploid number of chromosomes. Instead, it would have one too many or one too few chromosomes (passed on from the defective gamete).

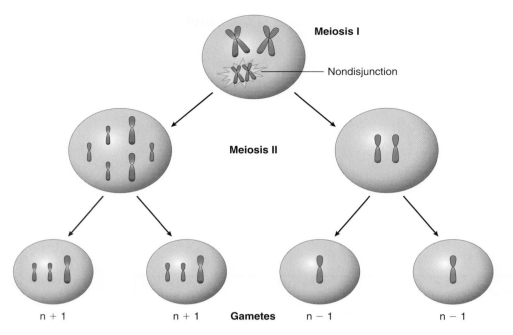

FIGURE 8-3 Nondisjunction During Meiosis I

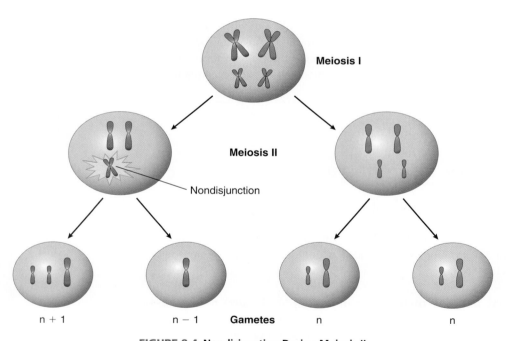

FIGURE 8-4 Nondisjunction During Meiosis II

If an individual is born with an abnormal chromosome number, this can result in a variety of symptoms, depending on which homologous pair is affected. An extra chromosome in one of the homologous pairs is known as a **trisomy**.

One well-known disorder of this type is Down Syndrome, caused by an extra copy of autosome 21. Trisomy 21 is the most common abnormality of chromosome number in the

United States, affecting one out of every 700 children born. Medical problems associated with Down Syndrome include mental retardation, facial abnormalities, short stature, heart defects, and susceptibility to respiratory infections.

INQUIRY AND ANALYSIS

1. You're a genetic counselor and the following karyotype is presented for analysis (see **Figure 8-5**). Write a summary of your findings as directed by your instructor. In your report, include the following information:

   - gender of patient
   - description of any chromosomal abnormalities present in the karyotype
   - name of abnormality (if present)

    Note: If additional information is needed to complete your report, do some research in your textbook or on the Internet.

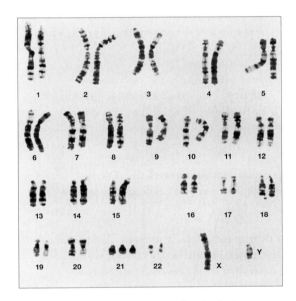

FIGURE 8-5 Karyotype for Analysis

2. Based on the gender of the person who provided the karyotype in **Figure 8-5**, diagram the most likely chromosome composition of their gametes.

3. Would it be possible for the person who provided the karyotype in **Figure 8-5** to have a daughter with Down Syndrome if he or she had a spouse with a normal karyotype? **Explain your answer.**

ACTIVITY 3 • RELATIONSHIP OF CHROMOSOMES, GENES, AND ALLELES

Not all genetic disorders are caused by an abnormal chromosome number. All of a person's traits, good and bad, are controlled by the **genes** that are located on your chromosomes. Since chromosomes exist in pairs, the genes on those chromosomes also are present in pairs (maternal and paternal).

Since you've inherited your genes from each parent, each pair of genes contains **one maternal and one paternal copy** of the gene. These copies are called **alleles.** In genetics, alleles are represented with letters of the alphabet. If **both alleles** of a pair have the **same genetic information,** the individual is said to be **homozygous** for that gene. If the two alleles are different, this is called **heterozygous.**

When an individual is **heterozygous,** it often occurs that **just one** of the alleles is expressed. An allele that exerts its effect even when paired with another, different allele, is referred to as **dominant.**

Recessive alleles are only expressed when **no dominant allele is present.** A dominant allele, in other words, **masks the expression of the recessive allele.** By custom, we indicate **dominant** alleles with capital letters (**R**) and **recessive** alleles with lowercase letters (**r**).

The combination of alleles for a trait is an individual's **genotype** (such as RR or Rr). The physical description of a specific trait is called the **phenotype** (such as blood type, hair color, or height).

INQUIRY AND ANALYSIS

1. Consider the following three genotypes for the production of the protein melanin. The **dominant allele (A)** produces normal skin color, but the **recessive allele (a)** inhibits melanin production, resulting in an albino (lack of pigment in the skin).

   AA   **Aa**   **aa**

   Which of the above genotypes represents an individual who is **homozygous** for the melanin trait? _____

2. Which of the genotypes listed in question one contains one maternal allele and one paternal allele? **Explain your answer.**

3. Which genotype(s) listed in question one represent(s) an individual that is **heterozygous**? ____

4. Would there be a **phenotypic** difference between AA and Aa when the genes are expressed? **Explain your answer.**

5. Which genotype listed in question one represents a person who is an albino? **Explain your answer.**

ACTIVITY 4 • DETERMINING INHERITANCE BY COMBINING ALLELES

When studying genetics, the most common question students ask is, "How can I tell which traits could be present in my baby?" To demonstrate how to find the answer to that question, consider the skin color trait discussed in **Activity 3.** Suppose we had two parents with normal skin color, both of whom have the albinism trait in their families. How can we calculate the probability that this couple could have an albino child (a **recessive** trait)? There are three simple steps you can follow to figure out the probability for the offspring.

INQUIRY AND ANALYSIS

Step 1 in the process is to determine all the possible genotypes and phenotypes for the melanin trait. Since we're working with two alleles per person, there are **only three** possible allele combinations.

1. Based on your knowledge of dominant and recessive alleles, fill in the phenotypes for each allele combination.

Genotypes	Phenotypes
AA	
Aa	
aa	

Step 2 in the process is to determine the genotype of each parent. With the completed genotype and phenotype information, we can determine the probability of this couple having an albino child. To simplify matters, assume that **both parents are heterozygous** for the melanin trait.

 Reminder: When gametes are formed during meiosis, the alleles of each homologous pair are separated into individual sex cells (eggs or sperm).

Since there is no way to know which egg and sperm will combine to form a child, all possible combinations have to be considered. An easy way to visualize these combinations is through the use of a **Punnett Square.**

Step 3 is to set up a Punnett Square for these two parents. The alleles for one of the parents are entered along the side of the square and the other parent on top.

It doesn't matter which parent is on which side of the square, as long as you line up the alleles with the proper rows and columns (as demonstrated in **Figure 8-6**).

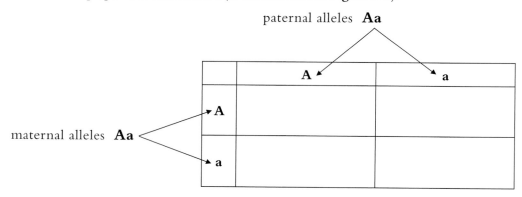

FIGURE 8-6 Constructing a Punnett Square

To complete the square (which will show you all the possible genotypes for children of this couple), you must combine the maternal and paternal alleles. If you combine one dominant maternal allele (A) with one of the paternal alleles (A), the resulting combination would be one possible genotype for their child (AA).

Each square in the Punnett Square represents a **25% probability** that this couple could have a child with that genotype. The four squares add up to **100%** (all the possible genetic combinations for this trait with this set of parents). **Each time a pregnancy occurs, the probability remains the same as shown in the square.**

	A	a
A	AA	
a		

To fill in the squares, combine the allele in the shaded box at the top of the column with the allele from the shaded box to the left of the row. The process is demonstrated in the first square above.

2. Combine the appropriate alleles to complete the other three boxes in the Punnett Square.

    What is the probability that this couple could have an albino child? _____

3. If this couple does have an albino child, what is the probability they could have a second albino child? **Explain your answer.**

4. If this couple has a child with normal skin color, is there any possibility that this normal child could have an albino baby? **Explain your answer.**

5. Assume that this couple has already had one albino child. The father says to you, with relief, that he's glad any future children he may have won't be albinos. Is he correct in his assumption? **Explain your answer.**

6. Steve has normal skin color. His Mom also has normal skin color, but his Dad is an albino. Steve marries Marilyn, who is an albino. What are the genotypes of all the individuals mentioned in the problem?

Note: In some cases the problem statement may not include enough information to determine a person's second allele. If so, enter a question mark (?) in place of the missing allele.

Genotypes:

Steve _____   Steve's Mom _____   Steve's Dad _____

Marilyn _____

7. What is the probability that Steve and Marilyn could have an albino child?

Hint: Don't forget to do Steps 1–3 in order!

Probability of an albino child _____

Probability of a child with normal skin color _____

ACTIVITY 5 • GENETIC TRAITS IN THE POPULATION

Can you roll your tongue as shown in **Figure 8-7**? Do you think the ability is inherited? As it turns out, determining whether traits such as tongue rolling have a genetic component isn't always as simple as it appears.

Research by the Human Genome Project has determined that, although tongue rolling is an inherited trait, it's not a simple matter of dominant and recessive alleles. If you do a little searching through the Internet, you'll find that other factors may be involved. For example, some studies showed cases of identical twins in which one twin was a tongue roller and the other wasn't! Clearly more research is needed to determine all the factors involved in the inheritance of this trait.

One way that scientists are investigating this question is by examining the frequency of this trait in the general population. To examine the frequency of this trait on a smaller scale, you'll sample family, friends, and acquaintances.

FIGURE 8-7 Tongue Rolling

INQUIRY AND ANALYSIS

1. Ask **20 people** the following question: "Can you roll your tongue into a tube?" Ask them to show you.

   Record the **number of people who can and the number of people who can't** roll their tongues in **Table 8-1**.

    Note: It's not necessary to indicate the numbers of males or females or adults or children; however, if you sample any **related individuals,** you should include that information when you post your results.

TABLE 8-1	Tongue Rolling Results	
		Totals
Number of Tongue Rollers (related to each other)		
Number of Tongue Rollers (NOT related to each other)		
Number of Non-Rollers (related to each other)		
Number of Non-Rollers (NOT related to each other)		

2. **Post** your results in your online classroom according to the directions given by your instructor.

3. **Tabulate** the results posted by your fellow classmates. Answer the following questions, **in reference to the total class results.**

   What percentage of the sampled population were tongue rollers? _____

   What percentage of the tongue rollers were members of the same family? _____

   What percentage were not able to roll their tongues? _____

   What percentage of the non-rollers were members of the same family? _____

4. Were there any pairs of identical twins? If so, did they have the same results on the tongue rolling test? **Explain your answer.**

5. Were there any pairs of fraternal (not identical) twins? If so, did they have the same results on the tongue rolling test? **Explain your answer.**

6. If you sampled **2,000 additional people,** would you expect the observed percentages to change? **Explain your answer.**

7. Research suggests that factors in addition to inheritance are involved in tongue rolling ability. Suggest one factor that could possibly influence the expression of this trait and explain your reasoning.

Exercise 9 • Molecular Genetics

OBJECTIVES

After completing this exercise, you should be able to:

- explain and demonstrate the double helix structure of a DNA molecule
- demonstrate knowledge of the base pairing rule in DNA and RNA
- compare and contrast DNA and RNA
- list and explain the steps of protein synthesis and the role of each type of RNA in the process
- discuss the effects of mutations such as base substitutions, deletions, and insertions on the DNA code

SUPPLIES

Activity 1

SUPPLIES FROM LAB KIT

- None needed

HOUSEHOLD SUPPLIES

Gummi Bears, 1 package

Toothpicks (rounded), 1 box

Stick licorice (such as Twizzlers®)

ACTIVITY 1 • BUILDING A DNA MODEL

Most cells have a nucleus. Depending upon the species, there's a set number of chromosomes within this organelle. Each chromosome contains a **DNA** molecule and its associated proteins.

DNA stands for **deoxyribonucleic acid**. If the DNA were isolated from a cell and then stretched out, it would look like a spiral staircase. The spiral staircase has been called a **double helix.**

The staircase has rails on the outside that serve as the backbone of the structure, while the steps separate the two rails. The "steps" of a DNA molecule are composed of **nitrogenous bases.**

There are only **four different types of bases** found in DNA, and these bases pair up in a specific way (see **Figure 9-1**). The **base pairing rule** for DNA is as follows:

- adenine (A) always pairs with thymine (T) and
- cytosine (C) always pairs with guanine (G)

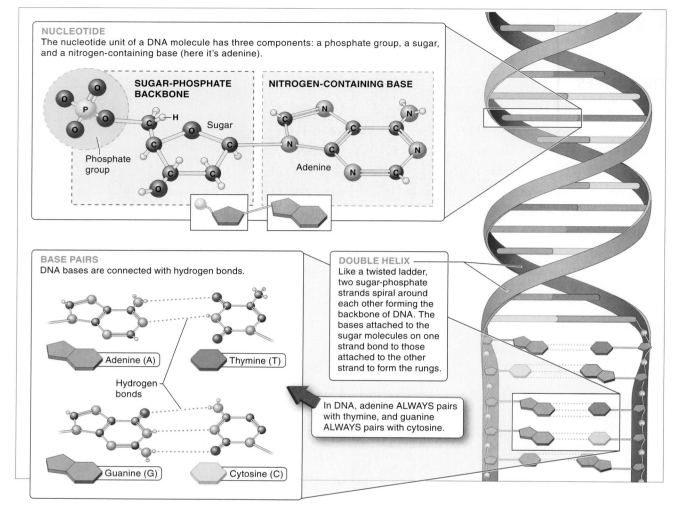

FIGURE 9-1 DNA Structure

In this activity, you'll build a model of DNA that simulates the three-dimensional structure of the molecule.

INQUIRY AND ANALYSIS

1. Take four different color Gummi bears and assign each color a DNA base. Fill in your color codes in **Table 9-1**.

TABLE 9-1 DNA Color Codes

Gummi Bear Color	DNA Base

2. Based on your color codes in **Table 9-1,** you'll build the following DNA sequence:

> T
> G
> A
> C
> G
> T
> G
> A
> C

Select a bear with the color that you have chosen to represent thymine. This will be the first base in your DNA sequence.

Slide the toothpick through the first bear (T) in a "paw to paw" direction.

Select a bear for the next base and repeat this process. Continue until you have a total of **nine toothpicks,** each with an appropriate **colored bear in the center.**

Note: Be sure to keep the toothpicks in the correct order (matching the stated DNA sequence).

3. Using the **base pairing rule**, slide a second bear onto each toothpick so that it's **to the right** of the first base.

   The colors of the new bears should represent the colors of the **complimentary bases** (according to the **base pairing rule**).

   Be sure to leave a small space between each base to represent the hydrogen bonding. **You've now created base pairs**.

4. Lay the licorice "**backbone**" along the **left side** of your toothpicks.

   Push the first base pair into the backbone near the **top edge** of the candy. Continue to add base pairs along the backbone, making sure they're evenly spaced.

   Attach the second DNA backbone to the other side of your exposed toothpicks. **Gently** twist your completed DNA model to form a helical structure.

5. Your DNA model contains a total of _____ bases.

6. The **DNA code** is a sequence of three-base units along one side of the DNA molecule. Units of three DNA bases are referred to as **triplets,** such as the three base unit **TGA.**

   How many triplets are present along the **left side** of the DNA molecule **you made**?
   _____

7. What is the **last triplet** on the **right side** of the DNA model? _____

8. If you were comparing your DNA to a classmate's DNA, would you use the same bases to construct their DNA molecules? _____ **Explain your answer.**

ACTIVITY 2 • TRANSCRIPTION

Imagine that you're having a house built. The architect draws up a set of blueprints for the builder. The builder, in turn, has to hire several subcontractors for the electrical, plumbing, and heating systems. The builder doesn't wish to give up the original set of blueprints, so copies are made for each of the subcontractors.

This same principle applies to a cell. The **nucleus** contains the **master set of blueprints,** the **DNA.** DNA molecules are needed to direct the activities that occur in the cytoplasm of each cell.

When other organelles need to manufacture materials for the cell, copies of the master blueprint are made, so that the originals are not damaged or lost. This process is referred to as **transcription.** The copies made by transcription will be in the form of another molecule called **messenger RNA (mRNA).**

There are several differences between DNA and mRNA:

- DNA is double stranded
- DNA contains the base thymine (T)
- DNA contains the sugar deoxyribose

- mRNA is single stranded
- mRNA contains the base uracil (U)
- mRNA contains the sugar ribose

In this activity you'll **transcribe a DNA code into mRNA.** The DNA sequence in this activity represents a gene located on one of the chromosomes in the nucleus of a cell that codes for a specific protein.

INQUIRY AND ANALYSIS

1. Since thymine isn't found in mRNA, the base pairing rule you learned in the previous activity must be altered. Complete **Table 9-2** by entering the appropriate bases.

TABLE 9-2 mRNA Base Pairing

DNA Base	mRNA Base
A	
C	
T	
G	

2. Looking at **Figure 9-2,** you can see that column one contains the DNA sequence found along **one side** of a double helix.

   Fill in the appropriate **mRNA sequence** in the **second** column.

   Reminder: mRNA doesn't contain thymine.

3. The mRNA code is composed of a **sequence of three bases** along the mRNA molecule. The three bases are referred to as a **codon.**

   ATG is the first **triplet** on the **DNA** sequence. What is the corresponding **codon**?
   _____

4. How many codons does your mRNA strand contain? _____

5. What is the base sequence of the **fourth** codon of your mRNA? _____

1 DNA	2 mRNA	3 Amino Acids
A		
T		
G		
T		
A		
T		
G		
T		
T		
T		
G		
A		
C		
G		
G		
G		
A		
G		
A		
C		
C		
C		
C		

FIGURE 9-2 Transcription and Translation

ACTIVITY 3 • TRANSLATION AND PROTEIN SYNTHESIS

Having copied the DNA into mRNA, the mRNA leaves the nucleus of the cell and travels to a ribosome.

 Reminder: Ribosomes are the sites of protein synthesis within a cell.

Proteins are composed of chains of **amino acids.** A chain of amino acids is also referred to as a **polypeptide chain** because the amino acids are held together with a type of covalent bond called a **peptide bond.**

Amino acids are brought to the ribosome by another type of RNA called **transfer RNA (tRNA).** The codons in the mRNA **determine the sequence and number of amino acids** being bonded into the corresponding protein.

INQUIRY AND ANALYSIS

1. Refer to the **mRNA codons** in **Table 9-3** to determine the amino acid sequence for your protein.

   To do this, just match each codon in your mRNA sequence to the corresponding amino acid as listed in the table.

   Fill in column three of **Figure 9-2** with the correct amino acids.

2. How many amino acids does your protein contain? _____

3. What is the **sixth** amino acid in the polypeptide chain? _____

4. If the sixth triplet read **GAA instead of GGA,** would the sequence of amino acids change? **Explain your answer.**

5. If the sixth triplet read **GGC instead of GGA,** would the sequence of amino acids change? **Explain your answer.**

6. What is the **name** of the bonds that attach the amino acids together within the **polypeptide** chain?

TABLE 9-3 mRNA Codons and Their Corresponding Amino Acids

codon	amino acid	codon	amino acid	codon	amino acid	codon	amino acid
AAU	asparagine	CAU	histidine	GAU	aspartic acid	UAU	tyrosine
AAC	asparagine	CAC	histidine	GAC	aspartic acid	UAC	tyrosine
AAA	lysine	CAA	glutamine	GAA	glutamic acid	UAA	terminator
AAG	lysine	CAG	glutamine	GAG	glutamic acid	UAG	terminator
ACU	threonine	CCU	proline	GCU	alanine	UCU	serine
ACC	threonine	CCC	proline	GCC	alanine	UCC	serine
ACA	threonine	CCA	proline	GCA	alanine	UCA	serine
ACG	threonine	CCG	proline	GCG	alanine	UCG	serine
AGU	serine	CGU	arginine	GGU	glycine	UGU	cysteine
AGC	serine	CGC	arginine	GGC	glycine	UGC	cysteine
AGA	argenine	CGA	arginine	GGA	glycine	UGA	terminator
AGG	argenine	CGG	arginine	GGG	glycine	UGG	tryptophan
AUU	isoleucine	CUU	leucine	GUU	valine	UUU	phenylalanine
AUC	isoleucine	CUC	leucine	GUC	valine	UUC	phenylalanine
AUA	isoleucine	CUA	leucine	GUA	valine	UUA	leucine
AUG	methionine (start)	CUG	leucine	GUG	valine	UUG	leucine

ACTIVITY 4 • INTERPRETING THE DNA CODE

You've seen that DNA triplets code for specific amino acids and that the amino acids must be attached in the correct sequence to form a functional protein (**Figure 9-3**).

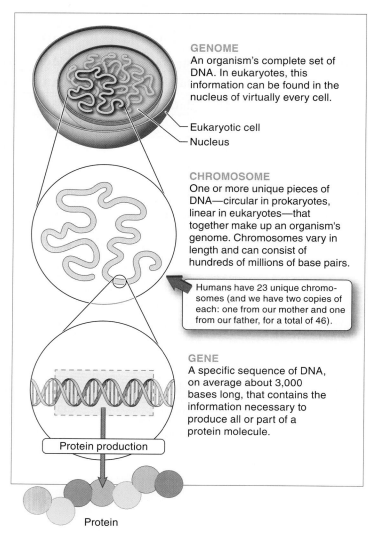

FIGURE 9-3 **From DNA to Proteins**

Now you can explore what happens if changes in the DNA triplets occur. These changes can occur from exposure to radiation, certain chemicals, and other environmental factors. In addition, errors can be incorporated into the DNA when the chromosomes are duplicated in preparation for mitosis and meiosis.

Changes in the DNA code are known as **mutations.** The effects of a mutation on phenotype can vary, depending on the type and location of the mutation within a gene. In this activity, you'll simulate the effects of various types of mutations.

INQUIRY AND ANALYSIS

1. Copy and paste the following section of DNA code into a word processing file. Your instructor will direct you to a **web location** where you can find the code to copy.

   CAGAATTTGATTGCCAATGCCACGAGTAACATTGCCAAAAATTGTCA

2. **Draw lines** between the triplets and **number** them in order from left to right.

   How many triplets are in this DNA code? _____

3. **Simulation** Go to: www.whfreeman.com/bres

   You'll find a simulation created for this activity. Follow the onscreen instructions to complete the activity.

   Click on Exercise 9 (Molecular Genetics).

   Click on Activity 4 (Interpreting the DNA Code).

4. Paste your DNA code into the second box on the screen.

   Click "Decode."

   The decoded sequence reads:

   The first type of mutation you'll investigate is called a **base substitution.** In a base substitution, one base is simply replaced by another in a specific section of the code.

5. **Replace the T in triplet number 6 with an A. Copy and paste** the revised code into the DNA decoder. Has the decoded sequence changed? If so, how?

   Additions and deletions are two other types of mutations that can occur. In additions, an extra base is inserted into a region of the DNA code. A deletion is the removal of a base.

6. Copy the **original, unaltered DNA strand** from step 1.

    Simulation

   Return to the simulation and **delete one of the A's in triplet number 6. Copy and paste** the revised code in the DNA decoder.

   Has the decoded sequence changed? If so, how?

ACTIVITY 4 • INTERPRETING THE DNA CODE 111

7. Both changes in the DNA code (the substitution and the deletion) were made in **triplet 6.** Which mutation caused the greatest alteration in the decoded sequence? **Explain** why one mutation caused a bigger problem than the other.

8. If the DNA mutation was an **addition,** would the errors in the code be more similar to those you observed in the base substitution or the deletion? Why?

 Biotechnology Today

Have you ever wondered where some of today's drugs come from? Would you believe goat milk? Researchers have developed a new technique for manufacturing drugs that involves genetically modified goats.

The first such drug approved by the FDA is being used to treat patients with a deficiency of the protein antithrombin (used by the body to dissolve blood clots). Patients with this problem are at high risk during surgery and childbirth, so the drugs could be administered as needed in the hospital. One of every 5,000 Americans has this disorder.

Milk is ideally suited for drug production because the mammary gland already manufactures proteins, so manufacturing a new type of protein isn't a big change in function. Scientists have been able to insert the additional code for making antithrombin into the chromosomes that carry the genes for milk production. The technique of producing drugs in animal milk is faster and less expensive than obtaining the protein from human plasma and doesn't hurt the goats at all.

Exercise 10 • Biology in Forensic Investigations

OBJECTIVES

After completing this exercise, you should be able to:

- critically observe and describe a crime scene
- compare and contrast fingerprint patterns and details
- perform a latent fingerprint analysis and interpret the results
- describe the role of DNA analysis in forensic investigations
- analyze and interpret the information in an STR profile
- draw conclusions and solve a crime using different types of forensic evidence

SUPPLIES

Activity 1
- a partner

Activity 2
SUPPLIES FROM LAB KIT
- stamp pad, 1
- magnifying glass, 1

ACTIVITY 1 • POWERS OF OBSERVATION

Many important scientific findings begin with a simple observation. Observation is one of the most important skills a scientist can have. Forensic investigators rely on evidence from their senses to perform their jobs effectively.

INQUIRY AND ANALYSIS

1. Go to: www.whfreeman.com/bres

   You'll find a simulation created for this activity. Follow the onscreen directions to complete the activity.

   Click on **Exercise 10** (Biology in Forensic Investigations).

   Click on **Activity 1** (Powers of Observation).

   After logging in to the simulation, your computer screen will show you two photographs of the same scene. One photo is the original. The other has several alterations in the appearance and positioning of items in the room.

    Note: You'll be comparing your skill with that of a partner. For this reason, it's essential that your partner isn't looking on while you locate the differences in the photos.

2. Visually compare the two images and find the differences. There are **12 differences** between the two photos.

   When you click the "start" button, you'll have **10 minutes** to find as many differences as you can.

   Each time you find a difference, click on that spot in the picture. You can click on differences in either of the two photos.

   Note: If your choice is correct, a red circle will appear around the difference to confirm that your choice is correct. If not, no circle will appear.

3. Refer to **Table 10-1** to determine your observation skill level.

   What is your skill rating? _____

TABLE 10-1 Observation Skill Ratings

Number of Differences Found	Skill Rating
11–12	expert
9–10	advanced
7–8	average
5–6	moderate
3–4	fair
0–2	needs improvement

4. Repeat the observation activity using your partner as the test subject. (Don't give them any hints!)

5. Refer to **Table 10-1** to determine your partner's observation skill level.

   What is his/her skill rating? _____

   Was your skill rating the same as that of your partner? _____

6. Did you both identify exactly the same alterations in the photos? If not, what were the differences?

114 EXERCISE 10 • BIOLOGY IN FORENSIC INVESTIGATIONS

7. Based on the results of this activity, why might it be important to have more than one person look at a crime scene?

8. What types of activities typically performed by a forensic investigator depend on accurate observation skills? Give **two examples** and **explain** how observation skills are involved in each example you cite.

ACTIVITY 2 • FINGERPRINT ANALYSIS

Formation of fingerprints

One of the most important functions of forensic science is to establish the identity of an individual. Fingerprinting, the oldest tool for human identification, has been used for centuries as a means of differentiating among people. In the late 1800s, Sir Frances Galton performed the first serious studies of fingerprints. Based on decades of fingerprint identification, no two people have as yet been shown to have identical fingerprint patterns; therefore, fingerprinting is a reliable means of **unique** identification.

The ridges of the **epidermis** form patterns. Everyone has a unique pattern of these ridges. The patterns are formed due to the shape of **dermal papillae,** which form a bumpy boundary layer between the epidermis and the dermis (**Figure 10-1**). The shape of the boundary layer determines the pattern of ridges and depressions we know as fingerprints. Ridge patterns are formed as a fetus and remain the same throughout your life.

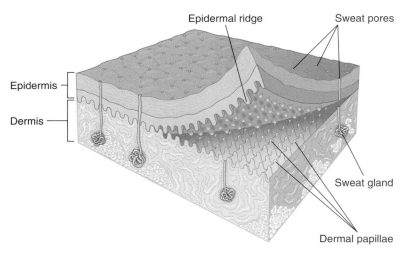

FIGURE 10-1 Dermal Papillae in the Skin Form Fingerprints

Basic fingerprint patterns

There are three basic fingerprint patterns that are most common in the population—**loops, whorls,** and **arches.** The fingerprint pattern for an individual finger is defined as the largest (most noticeable) feature on the finger pad.

Loop patterns (**Figure 10-2**) have ridge lines that enter and exit from the **same side** of the finger tip. A loop can face either to the left or right, in comparison to the body midline. A **double loop** has two separate loops within one print area.

(a) Loop (b) Double Loop

FIGURE 10-2 Loop Patterns

Whorls (**Figure 10-3**) are formed by a series of concentric ridges within the pattern area. Arches (**Figure 10-3**) have ridge lines that enter and exit from the **opposite sides** of the finger tip. **Simple arches** form a smooth curve at the top of the arch, while **tented arches** form a point within the arch.

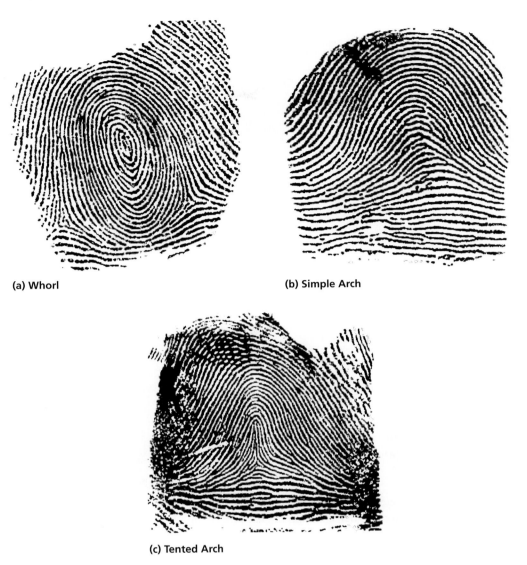

(a) Whorl

(b) Simple Arch

(c) Tented Arch

FIGURE 10-3 Whorls and Arches

Latent and known fingerprints

The surface of your fingers has a pattern of ridges formed by the dermal papillae. The toes, palms, and soles of the feet also have similar ridge patterns. Invisible impressions made by the finger ridges when you touch a surface are called **latent fingerprints.** Since latent fingerprints are invisible, they're often left behind accidentally and form part of the evidence used to solve crimes.

Known fingerprints are collected from a specific individual to serve as reference samples. Known prints can be collected with ink and paper or through electronic scans.

INQUIRY AND ANALYSIS

1. Suggest some other parts of the body that could leave prints.

   a.

   b.

   c.

2. Use the following instructions to take your own fingerprints and place them in the appropriate locations in **Figure 10-4**.

   Starting with the right hand, print the thumb first, then proceed toward the little finger moving from left to right. Place each print in the appropriately labeled boxes of **Figure 10-4**.

   To make the prints, place one side of each finger (**one at a time**) on the ink pad (from your lab kit). Roll the finger from one side to the other, covering the flat surface (up to the joint) with ink.

   Caution! Be careful with the ink in the pad. It can stain your clothing.

   Lightly, repeat the rolling motion to transfer the inked print onto **Figure 10-4**.

   Note: One of the most common problems with taking fingerprints is using too much ink, so be sparing. Also, if you press too hard on the paper, the print will smear and be difficult to read. You may want to practice on a plain piece of paper first.

   If the ridges aren't clearly defined, you've probably used too much ink. Just make another print without re-inking your finger. If the prints appear smudged, you've probably pressed too hard. Just try again, but roll you finger lightly over the paper.

3. Repeat the instructions in step #2 with the left hand.

Right Hand					
	Thumb	Index Finger	Middle Finger	Ring Finger	Little Finger

Left Hand					
	Thumb	Index Finger	Middle Finger	Ring Finger	Little Finger

FIGURE 10-4 Your Fingerprints

4. Examine each print and identify the pattern (using the information in **Figures 10-2 and 10-3**).

   Write the identification for each finger **directly below** the appropriate fingerprint in **Figure 10-4**.

   Note: If one of your fingerprints doesn't have the general characteristics associated with the definitions for the loop, whorl, or arch, your print is classified as **accidental.** Accidentals are the general classification for prints that don't fall into one of the three common print categories.

Fingerprint details

The broad categories of fingerprint patterns are relatively easy to see, but the **individual details within** the prints can be more difficult to identify. It's the type, number, and location of these individual details that make a fingerprint unique.

INQUIRY AND ANALYSIS

1. Return to your fingerprints in **Figure 10-4.** Identify two fingers that have the **same** basic fingerprint patterns (e.g., both whorls or both simple loops).

   Are the prints identical? **Explain your answer.**

2. Some examples of the types of details that make fingerprints different from each other are listed below:

   - **Bifurcation:** a point where a ridge splits in two (appears similar to a fork in the road)
   - **Ridge ending:** the point at which a particular ridge ends
   - **Divergence:** two parallel ridge lines diverge apart
   - **Delta:** area where two ridge lines merge into a point
   - **Scars:** healed injuries that appear as white lines on the print

   Examples of how to recognize these features in a fingerprint are shown in **Figure 10-5.**

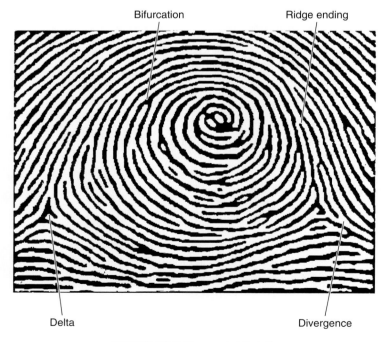

FIGURE 10-5 Fingerprint Details

3. Return to your fingerprints in **Figure 10-4. Highlight or circle** one example of each of the fingerprint details listed in #2 above. Use the magnifying glass from your Lab Kit to obtain a clearer view of the fine details within your fingerprints.

ACTIVITY 3 • DNA ANALYSIS

Anyone who watches crime shows on television is aware that it's possible to analyze DNA evidence found at a crime scene and match it to a specific person. DNA evidence can be used to identify the victim or the perpetrators of a crime and to exclude other potential suspects. In addition to convicting criminals, forensic DNA analysis has been used recently to overturn criminal convictions and release innocent people from jail (some after many years of incarceration).

In very general terms, each gene on a chromosome has a "genetic address" called a **locus.** Chromosomes also contain sections of **non-coding DNA.** The position of a particular segment of non-coding DNA on a chromosome is also known as a locus.

A DNA profile contains two pieces of information (**alleles**) for each locus. One allele is inherited from your mother and one from your father. During forensic testing, **several loci** (plural of locus) are used. By examining the DNA at several locations, the forensic scientist can gather more information to establish the source of a sample. The **more loci that are included in the analysis**, the more reliably you can match a DNA profile to a specific person, even though there may be other people in the population with similar DNA.

When you see a picture of a DNA profile in the media, you're most accustomed to seeing it portrayed as a DNA banding pattern (**Figure 10-6a**). However, a newer STR system is currently in use that employs computer software to display similar information to that in the banding pattern, but in a more readable format (**Figure 10-6b**), which you'll learn to read and analyze in this activity.

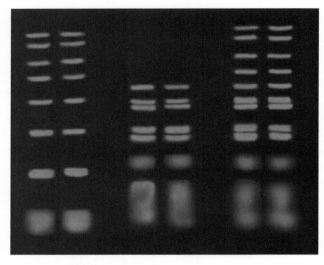

(a) DNA Banding

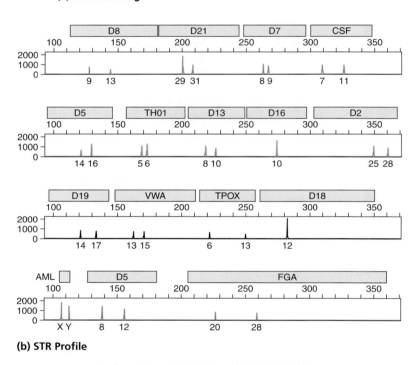

(b) STR Profile

FIGURE 10-6 Ways to Display DNA Profiles

INQUIRY AND ANALYSIS

1. What types of cells or tissues can be used as a **source** of the DNA for analysis? List several.

122 EXERCISE 10 • BIOLOGY IN FORENSIC INVESTIGATIONS

2. Go to: www.whfreeman.com/bres

    You'll find a simulation created for this activity. Follow the onscreen directions to complete the activity.

    Click on **Exercise 10** (Biology in Forensic Investigations).

    Click on **Activity 3** (DNA Analysis).

    Re-examine the living room photo you used in **Activity 1.** If this room was a crime scene, what types of evidence might be collected as a source for DNA analysis? **Explain your answer.**

Short tandem repeat (STR) analysis

DNA molecules are polymers of nucleotides. The DNA code within a gene consists of a series of nucleotide bases. The DNA used in forensic analysis is different from the DNA you studied in genetics because all the loci used for analysis are **non-coding loci.** In other words, they **don't** code for the synthesis of proteins.

An **STR** is a **small sequence of nucleotide bases that is repeated several times.** Each STR locus has two alleles. These alleles are inherited from your parents, as part of the DNA you received from the egg and sperm. So, if an individual had one allele that contained **14 repeats** of the STR sequence and a second allele that contained **17 repeats** of the same sequence, their **genotype** for this locus would be **14, 17.**

INQUIRY AND ANALYSIS

1. How is it possible that one parent could have 14 repeating segments at this STR locus, but the other has 17 repeats?

An STR analysis examines **multiple loci** and therefore produces an extremely informative DNA profile. Forensic DNA analysis in the United States makes use of the same **13 STR loci** that are tested by all forensic laboratories so that results will be consistent throughout the country. These results are stored in a computer database system called **CODIS (Combined DNA Index System)**.

 Reminder: The **more loci that are included in the analysis,** the more reliably you can match a DNA profile to a specific person, even though there may be other people in the population with similar DNA.

Each DNA sample is analyzed for multiple STR loci at one time. The 13 core CODIS loci are always included in this analysis. Additional loci may also be included to provide additional information. Each of the STR loci is designated with an abbreviation that consists of a combination of letters and numbers. The pattern of peaks from the STR samples is analyzed to determine the DNA "fingerprint" for a sample.

In addition to the STR loci, the DNA profile includes the AML (amelogenin) locus for gender determination. Females are XX and males are XY.

Interpreting an STR profile

The peaks on an STR profile represent the alleles at each locus. The numbers below each peak in the graph (**Figure 10-7a**) are used to designate the number of repeats of the STR pattern in that particular allele. Together, these two numbers constitute the **genotype** for that particular STR locus. In this example, the genotype is **11, 15.**

At some loci, you may notice that there is **only one peak** (**Figure 10-7b**). This represents a person who has **two identical alleles** at that site. In this case, the genotype is **11,11**. Because the alleles inherited from both parents are the same, they print out as only one peak on the profile.

 Reminder: The number beneath each peak represents the number of times an STR sequence is repeated at that locus.

The **height of the peak** represents the **concentration of DNA in the collected sample**. It's **not related to the number of STR repeats** within the allele.

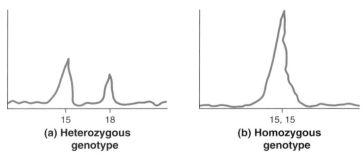

FIGURE 10-7 Two Sample STR Loci

INQUIRY AND ANALYSIS

1. In your own words, explain why some DNA loci show two separate peaks and others show only one peak.

2. Why is the peak height of the "11" allele in **Figure 10-7b** twice as tall as the peak height for the "11" allele in **Figure 10-7a**?

Analyzing a DNA profile is a two-step process. The first step is to match the alleles on the unknown profile with the known sample from a suspect or other individual. There must be a **100% allele match** or the suspect is excluded.

If the DNA profiles do match, a calculation is performed to determine the probability of seeing that exact profile in a particular population. Typically, these STR probability calculations result in probabilities rarer than one in a quadrillion (1/1,000,000,000,000,000). Based on this type of probability, you can state with confidence that a specific person is the source of the DNA in a particular sample.

3. It's easier to compare several STR profiles if you record and compile the genotypes into one location.

Mr. Smith is serving an extended prison term for a rape and murder in 1996. He has always maintained his innocence. Using modern DNA analysis techniques, semen evidence from the crime scene has been retested. The results of the analysis are shown in **Figures 10-8** and **10-9**.

Record the genotype for each locus in **Tables 10-2** and **10-3**.

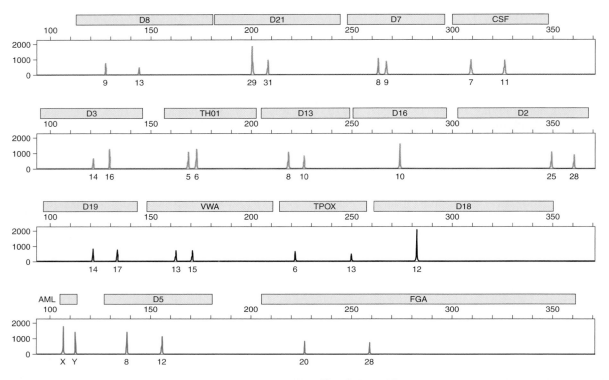

FIGURE 10-8 STR Profile of Mr. Smith

Note: This is an artificially generated DNA profile. Any resemblance to an actual person is purely coincidental.

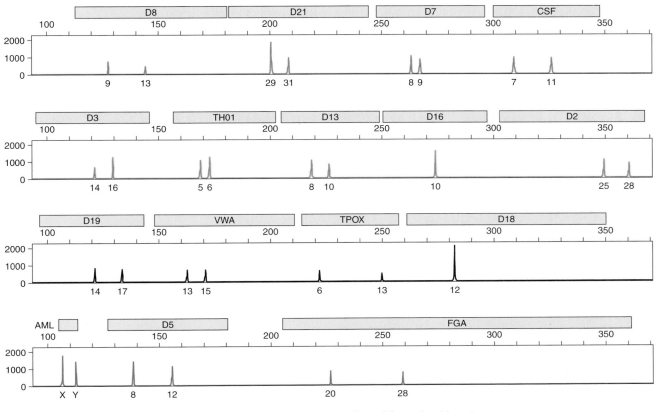

FIGURE 10-9 STR Profile of Semen Collected from the Crime Scene

Note: This is an artificially generated DNA profile. Any resemblance to an actual person is purely coincidental.

TABLE 10-2 Genotypes from Mr. Smith's Profile

	D8	D21	D7	CSF	D3	TH01	D13	D16	D2	D19
genotypes										

	vWA	TPOX	D18	AML	D5	FGA
genotypes						

TABLE 10-3 Genotypes from Semen Evidence

	D8	D21	D7	CSF	D3	TH01	D13	D16	D2	D19
genotypes										

	vWA	TPOX	D18	AML	D5	FGA
genotypes						

4. Based on the analysis of Mr. Smiths' DNA, would you advise his lawyer to seek a new trial? **Explain your answer.**

5. In regard to the analysis of Mr. Smiths' DNA, what would you conclude if the alleles in the **FGA** locus in **Table 10-3** were **25, 28**? **Explain your answer.**

Biotechnology Today

DNA technology is coming to a sports arena near you. Did you know that some sports memorabilia are authenticated with DNA? The ink used to designate official Olympic souvenirs is mixed with DNA from participating athletes at the Games. Superbowl footballs are also marked with invisible, but permanent, DNA. The DNA markings are unique and can be verified with a specially calibrated laser at any time in the future. These DNA tags act like a molecular "fingerprint" to stem the growing epidemic of sports memorabilia fraud.

ACTIVITY 4 • CRIME SCENE INVESTIGATION

The crime

Michael Jackson's estate has made arrangements to liquidate some of his assets by auction. The featured item is his famous sequined glove—bids starting at $25,000. However, the night before the auction, there was a break-in at the auction house—the famous glove was stolen from its glass display case!

The night of the crime, the video camera was turned off at 5:00 PM and remained off until the following morning. Two people were recorded entering the building, but there was no footage of them leaving.

Preliminary investigation led to the arrest of two suspects: a female employee of the auction house (Alice Nelson) and her brother-in-law (Mike Alvarez). The suspects were fingerprinted, interviewed, and DNA samples were collected. The auction house owner (Vincent Li) was also interviewed. Warrants were issued to collect the clothing the suspects wore the night of the crime and to search their vehicles.

You're a member of the forensics team investigating the crime. The following evidence was collected at the crime scene:

Near the exhibit:

- a white cotton glove
- drops of blood from the floor
- fingerprints from the electronic alarm key pad
- fingerprints from the broken glass on the floor
- fingerprints from the display case

Outside the auction house:

- a lug wrench found in a dumpster
- fingerprints from the lug wrench
- shoe prints in the mud near the dumpster

Your first job is to compare the fingerprints and DNA profiles of the suspects to evidence collected at the crime scene. Then, you'll use the results of your analyses combined with information from other members of the forensic team to determine whether the prosecutor should drop the charges against any of the suspects or hold them for trial.

INQUIRY AND ANALYSIS

The evidence

1. Go to: www.whfreeman.com/bres

   You'll find a simulation created for this activity. Follow the onscreen directions to complete the activity.

   Click on **Exercise 10** (Biology in Forensic Investigations).

   Click on **Activity 4** (Crime Scene Investigation).

2. Click on "**Crime Scene Photos**" to view photos taken of the scene before and after the crime and to obtain the fingerprints and DNA evidence. Click on "**Suspect Interviews**" to hear the statements given by Ms. Nelson and Mr. Alvarez. Click on "**Auction House Owner Interview**" to hear what Mr. Li told the police.

3. Click on "**Fingerprint Evidence**" to view the latent prints that were lifted from the electronic alarm key pad, the broken glass, and the display case.

   You'll also find files that contain the fingerprints taken from the suspects when they were arrested and those contributed voluntarily by Mr. Li.

   To download the fingerprints and DNA evidence to your computer, follow the directions on the screen.

4. Compare the fingerprints found at the scene with the prints collected from the suspects and the auction house owner.

   Examine both the print patterns and the fingerprint details (as you did in **Activity 2**).

   Examine and compare all prints from the crime scene to each person of interest. **Record your results in Tables 10-4 through 10-6.** For each match you find, **list all the features you used** to determine that the print was an **exact match** to that particular person.

TABLE 10-4 Fingerprint Comparison Alice Nelson

ID Number of Matching Print	Finger and Hand Matched	Explanation of Fingerprint Pattern and Matching Details

TABLE 10-5 Fingerprint Comparison Michael Alvarez

ID Number of Matching Print	Finger and Hand Matched	Explanation of Fingerprint Pattern and Matching Details

TABLE 10-6 Fingerprint Comparison Vincent Li

ID Number of Matching Print	Finger and Hand Matched	Explanation of Fingerprint Pattern and Matching Details

5. Click on "**STR Analysis**" to view the results of the **DNA tests** that were done on the **blood** and **epithelial cells** collected from the crime scene.

   You'll also find files that contain the STR analysis results of the samples contributed by the **suspects** (Ms. Nelson and Mr. Alvarez) and the **auction house owner** (Mr. Li).

6. Compare the STR profiles of all persons of interest with the blood and epithelial cells collected at the crime scene.

   Examine the results of the STR analyses (as you did in **Activity 3**).

7. Did any of the STR profiles match the known samples from the persons of interest? If so, which matches did you find? Explain the method you used to determine that the profiles matched.

8. Click on "**Additional Evidence**" to see the results of tests performed by other members of the forensic team.

   Summarize the findings of the other evidence teams in relation to each of the suspects.

   Alice Nelson:

Michael Alvarez:

Vincent Li:

INTERPRETING THE EVIDENCE

1. Was there sufficient forensic evidence against any of the persons of interest in this case for a prosecutor to charge them with the crime? Explain your answer in detail, including the logic and all the evidence you used to make your decision. Be specific.

2. To properly prepare for trial and convince the jury of this person's guilt, what other types of evidence would be useful? Give examples for each of the following:

   a. additional physical evidence:

   b. other evidence (not physical):

Exercise 11 • Evolution

OBJECTIVES

After completing this exercise, you should be able to:

- analyze the effect of pesticide applications on insect populations
- explain the concept of evolution by natural selection
- apply the principles of evolution by natural selection to real-life situations
- describe how beak structure is adapted to the type of food eaten by a bird species
- discuss how differential gene expression in an embryo can affect morphological features of an adult

Activity 2

SUPPLIES FROM LAB KIT

- None needed

HOUSEHOLD SUPPLIES

- assorted seeds, nuts, and other items (see detailed list in Activity 2)
- assorted tools (see detailed list in Activity 2)

large flat plates (dinner size), 2

measuring spoons and cups, 1 set

small zip-top bag, 1

kitchen timer (or watch or clock with a second hand), 1

ACTIVITY 1 • SELECTION PRESSURES AFFECT FUTURE GENERATIONS

Evolution refers to the changes that occur in populations over a series of generations. Those changes may occur quickly or extremely slowly and may be triggered by a variety of mechanisms, but all living organisms evolve.

The first method of evolution that was understood and explored was **natural selection.** Natural selection only affects heritable traits. Although actions you take in your daily life can affect your health and life span (negatively—such as smoking, or positively—such as choosing a healthy diet) and environmental factors can affect gene expression, only heritable traits can be passed onto your offspring. The combined genetic "library" for a population is referred to as the **gene pool.**

As a population expands, resources that are used by individuals (such as food, water, nesting sites) eventually become limited. Some individuals compete more effectively for scarce resources, survive longer, and provide more offspring to the next generation. The offspring that inherit the favorable traits of their parents will also compete and reproduce more effectively. Gradually, the favorable traits will spread through the gene pool. This modification of the gene pool is evolution by natural selection.

Populations evolve as a result of selective pressure from changes in their environment. Human activities, often inadvertently, alter environmental conditions and apply selection pressures in unexpected ways.

In this activity, you'll explore an unexpected "side effect" of a pesticide application in agriculture; it triggered changes in the population of a sucking insect called the California red scale, which feeds on oranges (see **Figure 11-1**).

FIGURE 11-1 Red Scale Infested Orange

The orange trees were sprayed to control the population of an unrelated insect (not the California red scale) that was harming the trees. DDT is a broad-spectrum pesticide that affects a large variety of species, often destroying organisms that were not the target of the spraying. Because of injuries to non-target species, DDT is no longer used as a pesticide in the United States, but is still in use elsewhere in the world.

DDT doesn't kill all pest species. Red scale insects, for example, are not significantly affected by DDT because they have hard, waxy protective coats. The graph in **Figure 11-2** shows the changes that occurred in the red scale populations over several years.

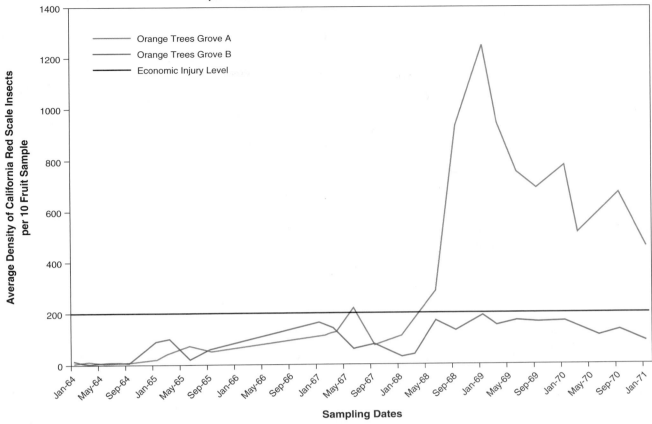

FIGURE 11-2 **Effect of DDT Application on Red Scale Insect Populations**

INQUIRY AND ANALYSIS

1. Based on the graph data, was the red scale insect a significant pest on orange trees in either grove between 1964 and 1966? **Explain your answer.**

2. If the red scale insects weren't a significant problem before 1966, suggest and explain some natural factors that might have been keeping the population under control.

3. Summarize the changes in the red scale population in Groves A and B beginning January 1967.

4. In what year do you think the farmers started using DDT on the orange grove? Why did you select that particular year?

5. Ladybird beetles (commonly called ladybugs) feed on scale insects and aphids. A ladybug can eat as many as 50 scale insects per day. In Grove B, the ladybug beetle population remained high and relatively constant, but in Grove A, the ladybug population varied over the six-year period. Based on the red scale population shown in **Figure 11-2, add a line to the graph** that represents the approximate population of ladybugs in Grove A over the six-year period. No specific numbers are given, but your graph should show the general trend you think the population took (population size rising and falling) over the six-year period.

6. Only one grove was sprayed with DDT. Was it Grove A or Grove B? **Explain your answer.**

7. How do the changes in Groves A and B provide an example of selection pressure acting on the red scale populations? What selection pressure(s) were involved? **Explain your answer.**

8. Assume that the initial DDT spraying killed 90% of the ladybugs. What new genetic characteristics would you expect to find in the ladybug population in Grove A between 1970 and 1975? **Explain your answer.**

9. You're the building manager of an apartment complex. You have a contract with a pest control company to spray the building for cockroaches every three months. Some residents in your building would like you to increase the frequency to weekly spraying. Based on the red scale insect case study, do you think this would be an effective approach to control the roaches in your complex? Why or why not? **Explain your answer.**

ACTIVITY 2 • EXPERIMENTS IN FORAGING

According to the principles of natural selection, there will be a **selective advantage** to physical characteristics that enhance survival by allowing animals to function better in a specific environment or help them to carry out necessary activities. One of the most important adaptations for food gathering in birds is the size and shape of the beak.

A species that can exploit a wide range of food resources is known as a **generalist.** Species that exploit only one or only a few food resources are known as **specialists.** Both strategies have advantages and disadvantages. For example a beak specialized for a particular type and size of seed will increase food gathering efficiency and therefore increase fitness. However, if environmental conditions change and that particular food decreases in abundance, a specialist could go hungry. The generalist, on the other hand, will be better able to exploit alternate, more available, food resources.

In this activity, you'll use a variety of "beak-like" tools to model the effect of beak shape on food gathering ability. Some beak types are highly specialized, whereas others can be used to collect a variety of foods.

INQUIRY AND ANALYSIS

1. Place two large flat plates (dinner size) on a table or counter. The "seed" list in **Table 11-1** is divided into three size categories: small, medium, and large. **Choose two items from each category** for your foraging experiment.

   Measure the specified amount for each item and place all the items in a small zip-top bag. Shake the bag well to mix the items and empty the bag onto one of the plates.

TABLE 11-1 Seed Choices

Small (1/4 Teaspoon for Each Choice)	Medium (1 Tablespoon for Each Choice)
uncooked rice	raisins
grass seeds	lentils (tiny)
shredded coconut	whole peppercorns
mini M&M's®	whole cloves
chocolate (or any color) sprinkles	dried cranberries
mustard seeds	split peas
fennel seeds	whole allspice
crushed red pepper (flakes)	pumpkin seeds
Grape Nuts cereal	granola cereal
tiny pasta (alphabet pasta, piccolini)	medium beads
oatmeal flakes	
candy Nerds	
small beads	
small bird seeds (millet, thistle seed)	

Large (1/8 Cup for Each Choice)

regular size M&M's®
leftover beans from seed experiment
sunflower seeds (with shell)
peanuts (without shell)
cashew nuts
pecans
crisp puffed cereal (Cocoa Puffs®, Trix, Cap'N Crunch)
dried macaroni
mini marshmallows
large beads
Gummi bears
Smarties

2. Select **two tools from each category** in **Table 11-2** (grasping and stabbing) to represent the beaks of various birds. Your responsibility is to **collect as many seeds as possible** during a **one-minute** foraging period.

   Select **one tool at a time** to complete your experiments.

   ### TABLE 11-2   Tool Choices

Grasping Tools	Stabbing Tools
tweezers	bamboo skewer (or toothpicks)
chop sticks	scissors
needle nose pliers	paring knife
broad nose pliers	fork
nutcracker	lobster pick
binder clips (small, medium, large)	mini screwdriver

3. Using the first tool, forage for seeds in the flat plate for **exactly one minute.** Transfer each seed you collect (individually) to the second plate, which represents "eaten" seeds.

    Note: You may only use **one hand** to manipulate the tool and transfer the seeds to the second plate!

   Sort and count the collected seeds. In **Table 11-3,** record the following information: **name of tool used, names of all seed types collected** (peanuts, M&M's®, etc.)**, and number of each seed type collected.**

   After you've recorded your data, return the seeds to the original plate.

4. **Repeat the procedure** twice more, for a total of **three foraging experiences** with the first tool.

   Continue with this procedure until you've tested all four tools (three trials each) **and recorded the data in Tables 11-3 through 11-6.**

   Calculate and record the average for each seed type over the three trials in **Tables 11-1 through 11-4.**

   Post your results as directed by your instructor.

TABLE 11-3	Results of Foraging Experiments					
Tool Name:	Types of Seeds Eaten					
Trial 1						
Trial 2						
Trial 3						
Average						

TABLE 11-4	Results of Foraging Experiments					
Tool Name:	Types of Seeds Eaten					
Trial 1						
Trial 2						
Trial 3						
Average						

TABLE 11-5	Results of Foraging Experiments					
Tool Name:	Types of Seeds Eaten					
Trial 1						
Trial 2						
Trial 3						
Average						

TABLE 11-6	Results of Foraging Experiments					
Tool Name:	Type of Seed Eaten					
Trial 1						
Trial 2						
Trial 3						
Average						

5. Based on your experimental results, describe any observed relationships between the tool type and the **number** of seeds eaten.

6. Which of your beaks functioned best as a "generalist" foraging tool? **Explain your answer.**

   Hint: Examine your collected data. Did the tool you selected "catch" an equal number of each food type or was it more successful with a specific type of food?

7. How do the results of your experiment illustrate the ways in which evolution by natural selection occurs? **Explain completely** using examples.

ACTIVITY 3 • PREDICTING DIET BASED ON BEAK SHAPE

In this activity, you'll extrapolate what a bird eats based on the structure of its beak.

INQUIRY AND ANALYSIS

1. In **Figures 11-3 and 11-4,** match each bird to its most probable food source, based on your evaluation of beak size and structure. Ignore the geographical availability of the food options for the various species.

   More than one species can eat the same food. A particular species can include **more than one type of food** in its diet. **Make all the possible matches.**

FIGURE 11-3 Birds with Various Beak Types

FIGURE 11-4 Potential Food Sources

2. **Record** your matches in **Table 11-7**.

TABLE 11-7	Matching Birds with Probable Diets
hummingbird	
zebra finch	
cardinal	
toucan	
macaw	
black legged stilt	
bald eagle	
white ibis	
goldfinch	

3. If you need help matching some of the birds with the most appropriate foods, check the information posted by your instructor to get matching hints and to see the correct matches.

4. How is the hummingbird beak style adapted for its food source? Explain how you made your matching decision.

5. Which birds in **Figure 11-3** are specialists? Which are generalists?

   Note: If you matched a bird to more than one type of food, list it as a generalist.

6. Ignoring geographic distribution, which birds from **Figure 11-3** do you expect could come to your yard to feed if you filled your bird feeder with:

   a. dried fruits:

   b. mixed nuts:

   c. sugar water:

   d. suet or peanut butter:

ACTIVITY 4 • REGULATORY GENES

All body cells are genetically identical—they're formed by the process of mitosis. If all body cells contain the same set of genes, how do they become specialized for different functions?

Although every cell has the same DNA, it's not expressing all of its genes at any given time. Cells contain a variety of control mechanisms that determine whether a particular gene will be **expressed** ("turned on") in a cell. For example, even though all body cells have the DNA directions for producing the protein hemoglobin, the gene is only expressed in red blood cells (and not in other cells around the body).

Variation in gene expression is one of the factors that leads to the evolution of new species. Changes on the molecular level lead to changes on the cellular level, which lead to changes in **morphology** (body structure). When enough changes accumulate, speciation can occur.

Consider the following example of how gene expression affects morphology. In chickens and ducks (and in humans), the foot begins development with webbing between the toes (see **Figures 11-5 and 11-6**). However, during foot development the cells that form the webbing die and the toes separate. Toe separation requires the activation of a **regulatory gene** (called BMP4) that instructs the cells between the toes to die (see **Figure 11-5**).

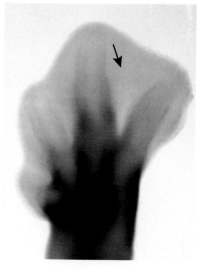  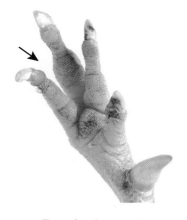

Inhibitor protein absent.

Arrow indicates pattern of cell death between toes.

Foot develops with separated toes.

FIGURE 11-5 Chicken Foot Development

In duck feet, however, an inhibitor protein prevents the expression of the BMP4 gene. The cells between the toes survive and the duck develops a webbed foot (see **Figure 11-4**).

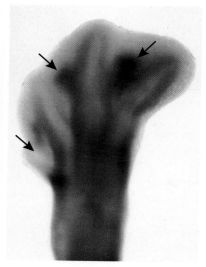 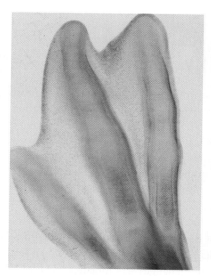 

Arrows indicate presence of inhibitor protein.

Absence of cell death between toes.

Foot retains webbing between toes.

FIGURE 11-6 Duck Foot Development

There are groups of genes that control the body plans of animals. This organizational pattern can be understood by thinking like a builder who's constructing a new house. The house is divided into "segments" (the layout of the rooms). In a similar fashion, most animals have sections of the body (called **segments**) that are organized in sequence from one end of the body to the other. In an insect, for example, there are three longitudinal segments: the head, the thorax, and the abdomen.

Each room of the house is constructed to accommodate specific structures and functions. For example, a kitchen needs plumbing for the sink, gas lines for the stove, and electrical lines for the lights. Other rooms have other requirements. How can a similar system work in a living organism?

In fruit flies, a group of regulatory genes called **homeobox (Hox) genes** make sure that appropriate body parts develop on the appropriate segments—for example, eyes and antennae on the head, legs on the thorax.

INQUIRY AND ANALYSIS

1. Go to: www.whfreeman.com/bres

   Click on **Exercise 11** (Evolution).

   You'll find a **Simulation** created for **Activity 4: Regulatory Genes**.

   Follow the on-screen instructions to complete the activity and record your answers below.

2. In the simulation, the **color coding** on the fruit fly chromosome and the body segments of the fruit fly embryo and adult shows **which Hox genes control the structures** that form on those segments.

   Observe the positions of the various Hox genes on the fruit fly chromosome and the corresponding control of the development of various body segments. How many Hox genes are on the fruit fly chromosome?

3. How many Hox genes control structures on the head? The thorax? The abdomen?

4. If we added another set of Hox genes to the fruit fly, would this probably result in an increase or decrease in the complexity of its body structure? **Explain your answer.**

5. Based on your gene and embryo observations, is there a correlation between the horizontal position of the Hox genes along the fruit fly chromosome and the head-tail axis of the fruit fly? **Explain your answer.**

6. Simulation

   Return to the simulation and color-code the mouse embryo in a similar fashion to that shown on the fruit fly as follows:

   Click on the first gene of the mouse chromosome. Next, **click on the body segment** that you think is controlled by that gene.

   Note: When you make the correct match between the genes and the body segments, the color of the gene will appear on the mouse embryo.

   Repeat the process until all the genes on the mouse chromosome have matching colors visible on the mouse embryo.

7. Is the relationship between the position of the Hox genes along the mouse chromosome **similar to or different than** that observed in the fruit fly? **Explain your answer.**

 Biotechnology Today

New research on fruit flies has shed light on the genes involved in alcoholism in humans. It turns out that alcohol affects fruit flies similarly to the way it affects humans. Scientists have been able to identify specific gene complexes related to alcohol sensitivity (alcoholics have a greater metabolic tolerance for alcohol consumption than the average person).

Interestingly, 23 of these genes have human equivalents, which researchers suggest could be linked to alcohol sensitivity in people. The new findings may also lead to the development of treatments for alcoholics.

ACTIVITY 5 • MANIPULATING GENE EXPRESSION

Even if genes aren't expressed, they're still part of an individual's genome. The genes are **conserved;** in other words, they remain part of the genome and form the genetic history of a species. This is especially noticeable in regulatory genes (such as the Hox complex), which have remained largely unchanged over the course of evolution.

The number of Hox genes has actually increased over time. When the DNA was copied to form gametes during meiosis, duplicate copies of the genes were accidentally produced. When Hox genes were duplicated, mutations in the duplicate copies allowed then to serve slightly different developmental functions. However, the basic code pattern of the Hox genes has been conserved and so we're able to see the evolution of related body structures through the animal kingdom. In general, the more copies of the Hox genes that are present in a particular animal group, the more complicated morphological structures appear. Body structures have evolved to become more complex and have taken on new functions over time.

Since genes are still present many generations later, if something activates gene expression, surprising physical characteristics can arise in a species. Gene expression is controlled by a **molecular tool kit** that consists not only of these highly conserved regulatory genes, but also of **transcription factors** and other types of **chemical signaling molecules.**

Transcription factors work by binding to specific DNA sequences and controlling transcription of mRNA. In other words, transcription factors can **either stimulate or block** the production of a particular protein from a gene.

In this activity, you'll form hypotheses about the effects of various "molecular tools" on the expression of genes and then perform experiments to determine whether your hypotheses were supported.

INQUIRY AND ANALYSIS

1. Go to: www.whfreeman.com/bres

Click on **Exercise 11** (Evolution).

You'll find a **Simulation** created for **Activity 5: Manipulating Gene Expression.**

Follow the on-screen instructions to complete the activity and record your answers below.

In the simulation, select fruit fly **Embryo 1.** Embryo 1 has a normal genetic makeup in regard to the formation of eyes.

The eye formation gene in Embryo 1 produces a **transcription factor called TF1,** which is necessary to activate other genes essential to eye formation.

 Note: The "eye formation" gene is responsible for activating a collection of genes that determine eye structure, such as single/compound lens formation, eye size, shape, etc.

2. **Transcription factor 2 (TF2)** blocks the expression of the eye formation gene.

   Form a hypothesis: What will happen if you block the activation of the **TF1**?

   Simulation

   Return to the simulation and select **transcription factor 2 (TF2)** from the molecular tool kit. Follow the on-screen instructions to test your hypothesis and watch the development of the fly from your altered embryo.

3. What effect did your molecular tool have on eye development?

4. A fruit fly embryo (**Embryo 2**) has a mutation of the eye formation gene so that the gene was permanently inactivated. Based on your results with Embryo 1, predict the effect that would have on eye development.

5. A research study has discovered that the *Pax6* gene in mice, which also controls eye development, is quite similar in gene sequence to the eye formation gene of the fruit fly.

   However, the single-lens eye of a mouse is quite different in structure and function from insect compound eyes, which have multiple lenses and form multiple images (see **Figure 11-7**).

FIGURE 11-7 Single Lens and Compound Eyes

6. **Circle your choice** in each of the following statements to form hypotheses about the effect of the mouse *Pax6* gene in the fruit fly embryo.

   Hypothesis 1: If I insert the mouse *Pax6* gene into the genome of Embryo 2, **eyes will form/eyes won't form.**

   Hypothesis 2: If eyes form in the fruit fly, they will have **single lens eyes/compound eyes.**

   Explain the reasoning you used to form your hypotheses.

7. Simulation

   Return to the simulation and select fruit fly **Embryo 2**.

   To test your hypotheses click on the **gene** icon from the molecular tool kit and select the ***Pax6*** gene. Follow the on-screen instructions to test your hypotheses.

8. What effect did the *Pax6* gene have on eye development?

   If eyes developed in your embryo, were they single lens or compound eyes?

9. If a compound eye formed in your embryo, what principle of evolutionary genetics was involved? **Explain your answer.**

10. **Simulation**

    Return to the simulation and select fruit fly **Embryo 3.** Embryo 3 has a normal eye formation gene.

    Form a hypothesis: What do you think would happen if you inserted the *Pax6* gene into the embryo **somewhere else on the body** (not on the head)?

11. **Simulation**

    Follow the on-screen directions to test your hypothesis and view the development of the fly from your altered embryo.

    What effect did the *Pax6* gene have in the location where it was placed?

12. **Simulation**

    Return to the simulation one more time and select fruit fly **Embryo 3** again. Follow the on-screen directions to manipulate the **Pax6** gene and the development of the fly from your altered embryo.

    What effect did the *Pax6* gene have in the multiple locations where it was placed?

13. What do you think would happen if *Pax6* was added to fruit fly **Embryo 1**? Explain your reasoning.

14. How is it possible that the mouse *Pax6* gene is expressed in the fruit fly. Explain in detail.

15. A new organism was discovered in a fossil bed. It appears to be directly related to an ancestral species, except that it has an additional body segment with an extra pair of legs. Could there be a connection between the development of the added body segment and Hox genes? **Explain your answer.**

16. Is the extra pair of legs likely to give the new organism any adaptive advantages? If not, why not? If so, give two examples.

Exercise 12 • Determining Your Ecological Footprint

OBJECTIVES

After completing this exercise, you should be able to:

- estimate your household energy and water usage
- analyze your usage patterns to determine ways to save energy and/or water in your household
- compare your water usage to national averages
- communicate your position on energy utilization issues to your elected representatives

SUPPLIES

SUPPLIES FROM LAB KIT

- None needed

Activity 2
HOUSEHOLD SUPPLIES

a bucket (should hold a couple of gallons), 1

measuring cups or other containers of known volume, 1 set

watch or clock with second hand, 1

ACTIVITY 1 • ENERGY AUDIT

Recently we've seen an increase in energy costs with regard to electricity and gasoline. Most experts feel that these increases will continue over the next several years as the demand for energy around the world grows. In this activity you'll assess the number of appliances in your home that consume energy and devise ways that you and your family can conserve energy.

INQUIRY AND ANALYSIS

1. Prepare an inventory for your home that includes **every household appliance** that uses energy. Examples include television sets and light fixtures, but not items that are used outside the home (such as lawn mowers). **Be complete.**

2. **Record** the collected information in **Tables 12-1** and **12-2**. For each item listed, enter the **number of similar items** in your home. For example, if you have 10 ceiling light fixtures, enter the number "10" in the column entitled "number of each" next to the light fixture row.

3. If you have items in your home that aren't listed in the tables, include them under **Additional Items.**

TABLE 12-1 Inventory of Energy Using Appliances in Your Home

Appliances	Number of each	Fixtures	Number of each
range (oven and cook top)		ceiling light fixtures (regardless of the number of bulbs in each)	
dishwasher		ceiling fan	
refrigerator		lamp	
freezer		night light	
furnace		cable TV boxes	
hot water heater		home security system	
air conditioner		free-standing fans	
sump pump		humidifier	
clothes washer		de-humidifier	
clothes dryer		air purifier/ionizer	
garbage disposal		space heater	
range exhaust hood		thermostat	
toaster		bathroom exhaust fan	
food processor			
blender			
mixer			
knife sharpener			
can opener			
toaster oven			
waffle maker			
coffee bean grinder			
microwave oven			
vacuum cleaner			
bread maker			
baby bottle warmer			
rice cooker			
electric wok			
sewing machine			
coffee pot			
electric frying pan			
crock pot			

TABLE 12-2 Inventory of Energy Using Appliances in Your Home (Continued)

Miscellaneous	Number of Each	Additional Items	Number of Each
cell phone			
cell phone battery charger			
battery charger			
radio			
clock			
television			
CD player			
video game console			
DVD player			
iPOD and accessories			
handheld video games			
computer and monitor			
printer			
power tools			
musical instruments			
paper shredder			
electric razor			
hair dryer			
curling iron or curlers			
cordless telephone			
telephone answering machine			

4. Based on your home inventory, list **five** suggestions for ways in which **your household** could decrease its energy use **without** major life-style changes. These must be suggestions that are relevant to **your** household specifically and related to your energy survey (not simply energy saving tips found on the Internet).

For each suggestion, explain **why** it's relevant to your household and **how** it will help you save energy. **Be complete.**

Submit your suggestions according to the directions given by your instructor.

a.

b.

c.

d.

e.

5. From the inventory you compiled in **Tables 12-1** and **12-2,** list five items that, **without causing a major inconvenience** to your life, you could **eliminate** in order to save energy. **Explain** your reasoning for each item.

a.

b.

c.

d.

e.

6. Motor vehicles are responsible for almost two-thirds of oil consumption in the United States. A typical American car burns over 800 gallons of gas per year. In that one year, a car can release up to 16,000 pounds of CO_2. How many miles you drive on those 800 gallons depends on the car's **fuel efficiency**—the number of miles per gallon (MPG).

   Use information from your studies of energy issues to convince someone to drive a car that gets better mileage. On the following page, **list and explain three different reasons** why driving a fuel efficient car is beneficial to **individuals,** the **community,** and/or the **nation.**

Submit your reasons according to the directions given by your instructor.

a.

b.

c.

 Biotechnology Today

Industrial Ecology is a new branch of environmental science that seeks to combine the needs of industry with those of the environment. Industrial ecologists look for innovative methods of removing wastes and using them to produce useful materials.

A new application for industrial ecology lies in the production of fuels from algae. Algae use photosynthesis to produce food molecules, and they grow amazingly fast. Excess energy is stored within the cells in the form of oil. An algal colony can double in weight several times a day and in the process, produce significant quantities of oil—15 times more oil per acre than other plants used for biofuels, such as corn or switchgrass. Some species of algae contain as much as 50% oil.

In addition, algae can grow under a variety of environmental conditions that plants can't tolerate, such as in salt or contaminated water. Growing algae for fuel can actually help clean up some serious environmental problems. One kilogram of algae removes 1.8 kilograms of carbon dioxide (an important contributor to global warming).

ACTIVITY 2 • WATER USAGE

As you know, **potable** (drinkable) water is a non-renewable resource. Many locations around the United States routinely experience water shortages and the duration of the shortages is increasing. Just like any other non-renewable resource, conservation measures are necessary. In this activity, you'll perform a three-day survey of the water usage in your home.

Read through the instructions completely before beginning your audit.

INQUIRY AND ANALYSIS

Showering

1. To determine how much water you use when showering, use the following procedures:
   - **Measure the volume** of the bucket. The bucket volume is _____ gallons.
   - Turn on the shower. Set the water pressure at the same level you use when taking a normal shower.
   - Place the bucket under the running water. Using a watch with a second hand, time how long it takes to fill the bucket.
   - Round the time to the nearest 5-second mark (for example, 13 seconds would be rounded to 15 seconds).

   It took _____ seconds for my bucket to fill with water.

2. Determine the **flow rate per minute** with the following formula:

   $$\frac{60 \text{ seconds}}{\text{\# of seconds to fill the bucket}} \times \text{volume of the bucket} = \underline{\quad} \text{ gallons/minute}$$

3. If you let the water run (for example, to get hot) before you step into the shower, record the number of minutes the water was running before each shower.

   gallons/minute × minutes water was running before shower began = _____ gallons consumed before shower

4. The equation in Step 2 tells you how many gallons of water you consume per minute of showering, but how much water do you use in your entire shower?

   To find out, multiply the number of gallons per minute by the number of minutes you spent in the shower each of the three days being audited.

   gallons consumed before shower + (gallons/minute × length of shower) = _____ gallons consumed in one shower

5. To determine the total volume of water used in showering, multiply the calculated shower volume by the number of showers taken on **each of the three days** being audited.

   > **Note:** Be sure to include **all** showers taken in your household—not just your showers.

   calculated shower volume × total # showers taken = _____ gallons consumed in showering

6. Record your showering totals for each of the three days in **Table 12-3**.

Toilet flushing

1. Take the lid off the toilet tank and look for the volume. It should be written inside the tank (in liters or gallons per flush). If not, look for the brand and model number of your toilet. You should be able to find the volume on the Internet.

2. To determine the total volume of water used in flushing, multiply the tank volume by the number of flushes on **each of the three days** being audited. Record your totals in **Table 12-3**.

   tank volume × total number of flushes = _____ gallons consumed in flushing

OTHER HOUSEHOLD WATER USES

1. To determine the amount of water used for other activities, you may need to devise some estimation strategies. **Devise ways to measure water use as accurately as possible,** using measuring cups or beverage containers to determine how much water you use per minute in activities such as dish washing, clothes washing, or washing your car.

2. Look for manufacturer's manuals and other sources (such as the county water department or the Internet) that may help you estimate the amount of water used by home appliances.

INQUIRY AND ANALYSIS

Completing the audit

1. Complete the **Water Usage Survey in Table 12-3** for **three typical days.** Include **both work and weekend days.** Be as accurate as possible.

2. Record the dates of your survey in the spaces below, as well as the number of people in your household.

   Dates of water use survey: Day 1 _____

   Day 2 _____

   Day 3 _____

   Number of people in household: _____

3. Calculate your household's **average three-day water consumption** from your results and enter the information in **Table 12-3**.

4. The average volume of water used **per person per day** in the United States is **70 gallons.** How does your water usage compare to the national average? **Explain your answer.**

TABLE 12-3 Water Usage Survey

(Round off your water consumption estimates to the nearest gallon.)

Domestic Water Use	Gallons Consumed (Day 1)	Gallons Consumed (Day 2)	Gallons Consumed (Day 3)
Washing hands/face			
Showering (water consumed before and during shower)			
Taking a bath			
Brushing teeth			
Washing food			
Cooking			
Drinking tap water			
Flushing the toilet			
Washing clothes by machine			
Hand washing clothes			
Washing dishes in dishwasher			
Hand washing dishes			
House cleaning			
Washing the car			
Watering the lawn			
Watering plants			

Other Uses (specify type of use):

Total Daily Consumption			

Average Three-Day Consumption =

5. **Over the course of several weeks,** what **three** activities would use the largest amount of water in your household? **Explain your answer.**

6. Make **five different practical** suggestions for reducing water use in **your** household. The suggestions **must not** require significant life-style changes. They **must** be suggestions related to **your particular household,** not just general suggestions copied from other sources.

   For each suggestion, explain **why** it's relevant to your household and **how** it will help you save water. **Be complete.**

   Submit your suggestions according to the directions given by your instructor.

   a.

   b.

c.

d.

e.

7. If you purchased a water heater that provided instant hot water for your shower, how much water would be saved per week in your household?

ACTIVITY 3 • GETTING INVOLVED

You've seen from the previous activities that it requires effort and thought to conserve energy and water. Imagine if you were able to share your ideas, not only with your classmates, but with a larger audience.

INQUIRY AND ANALYSIS

1. Draft a letter to your congressional representative (either House- or Senate-elected from **your** district) outlining **three proposals** that you recommend to significantly reduce **energy consumption** in the United States.

    The proposals should include actions that could be **easily implemented by a typical American household.**

    Make sure the letter is written in **YOUR OWN WORDS. Support your statements** with facts and examples based on your energy survey or other research you've done for this assignment.

2. **Submit a copy** of the letter to your instructor for approval. When your submission has been approved, email a copy of the letter to your congressperson. Visit www.house.gov or www.senate.gov to find the names of your elected representatives.

3. Submit confirmation that your email has been sent, as directed by your instructor.

Exercise 13 • Growth Patterns and Nutrient Transport in Plants

OBJECTIVES

After completing this exercise, you should be able to:

- explain the relationship between absorption of water by seeds, seed swelling, and germination
- explain the effect of seed orientation on direction of leaf, stem, and root growth
- describe the distribution and storage of energy reserves in plants
- describe an adaptation in plant roots that increases absorptive ability
- explain the process and identify the structures by which water and nutrients are transported through a plant stem
- predict the effect of environmental conditions on transpiration rate
- analyze tree ring growth and develop a history of climatic events

SUPPLIES

Activity 1

SUPPLIES FROM LAB KIT

- None needed

HOUSEHOLD SUPPLIES

dried lima beans or large kidney beans, 200

tap water, room temperature, ½ cup

containers that nest, clear (such as plastic cups), 8

waterproof marking pen (such as Sharpie), 1

metric ruler, 30 cm (12 inch), 1

kitchen timer (or watch or clock)

Activity 2

SUPPLIES FROM LAB KIT

- None needed

HOUSEHOLD SUPPLIES

- appropriate indicator for a starch test

dried lima beans or large kidney beans, 4

tap water, room temperature, ½ cup

containers that nest, clear (such as plastic cups), 2

waterproof marking pen (such as Sharpie), 1

measuring spoons, 1 set

paper towels, 1–2 sheets

Activity 3

- DVD: *Dissection of the Fetal Pig*

SUPPLIES FROM LAB KIT

- None needed

HOUSEHOLD SUPPLIES

potato, white, 1

celery stalks with leaves, fresh, 2

food coloring (preferably red or blue), 1

tap water, 1 cup

glass, tall enough to hold the stalks of celery, 1

ACTIVITY 1 • STRONG SEEDS

(Adapted from The Green Machine. CSIRO ANU Science Education Centre. Canberra, Australia. 1997)

Flowering plants reproduce by producing seeds. A **seed** is the plant equivalent of a fertilized egg. It contains an **embryo and stored food** to support the embryo's development, enclosed in a protective **seed coat.**

The embryo remains dormant within the seed until environmental conditions are favorable, then the seed will **germinate.** Germination is the process by which a seed resumes growth and development after its period of dormancy. In desert conditions, seeds can lie dormant for many years until sufficient rain falls to trigger germination.

The first step in germination is for the seed to **imbibe** (absorb) water. As the seed takes on more and more water, it begins to increase in volume and exerts pressure on the soil around it. The pressure breaks up the heavy soil above the seed, allowing the root and shoot to push through the soil. In addition, the pressure opens up air spaces around the seed, providing access to oxygen for cell respiration. This activity will demonstrate the first step of seed germination in the soil.

INQUIRY AND ANALYSIS

1. Place 20 dried beans in each of four clear plastic containers. Shake the beans in each container so that they lay as flat as possible, forming a level surface.

   - With a permanent waterproof marker, **label** the cups 1–4.
   - Add enough room temperature tap water to **just cover** the top of the beans.
   - Place a second, empty plastic cup inside each of the cups holding the beans.

2. Place the following number of beans in each of the top cups (see **Figure 13-1**):

   Cup 1 – empty
   Cup 2 – 10 beans
   Cup 3 – 20 beans
   Cup 4 – 40 beans

Cup 1 Cup 2 Cup 3 Cup 4

FIGURE 13-1 Strong Seed Setup

3. On each of the four setups, use a permanent waterproof marker to mark the position of the rim of the bottom cup on the side of the top cup (as shown in **Figure 13-2**).

FIGURE 13-2 Cup with Rim Position Marked

 Note: Mark the position of the rim of the empty cup (Cup 1) **without pushing down** on the beans.

4. **Every hour for the next six hours:**
   - Lift the top cup in each setup and check to make sure the beans are still covered with water. If not, add enough tap water to **just cover the beans** (and no extra).
   - Place the top cup back in each setup and **mark the position** of the rim of the bottom cup on the side of the top cup.

5. At the end of the six hours, **measure the difference** between the **first line** you drew and the **last line** to the **nearest tenth of a centimeter** (for example, 2.8 cm). **Record** your results in **Table 13-1**.

TABLE 13-1 Results of Strong Seed Experiment

Setup	Distance (centimeters)
Cup 1	
Cup 2	
Cup 3	
Cup 4	

6. Remove several beans from the various containers and compare their appearance with that of the dried beans. Do the soaked seeds differ in appearance from the seeds before they were soaked? If so, describe the differences.

7. If there were differences in the dried vs. soaked beans, what caused the differences you observed?

8. Was there a control in this experiment? If so, which setup was it? **Explain your answer.**

9. If you planted the beans you soaked in this experiment in your garden along with an equal number of dried beans directly from the package, predict which beans should sprout the most quickly. **Explain** the rationale you used to make your prediction.

10. Can germinating seeds lift half their weight?

Their own weight?

Double their weight?

Explain your answer. Support your explanation with facts from your experiment.

11. If a seed failed to imbibe water, do you think this would affect its ability to sprout successfully? **Explain your answer.**

12. The package of dried beans has directions for quick soaking. Here are the instructions: "Place the beans in a large pot. Add enough water to cover the beans by two inches. Bring the water to a boil. Boil for two minutes. Remove from heat, cover, cool, and let stand for one hour."

If you followed these instructions, then planted the soaked beans, would you expect them to sprout successfully? **If so, explain your answer. If not, why not?**

ACTIVITY 2 • RESPONSES TO ENVIRONMENTAL STIMULI

One of the basic characteristics of life is response to environmental stimuli. Plants are no exception. This response is referred to as a **tropism**. A tropism involves **directional growth** toward (**positive**) or away from (**negative**) some type of environmental stimulus.

We're all familiar with the fact that plants on your windowsill grow toward the light, but what about seeds? Can seeds exhibit tropisms? When a seed is buried underground, does the stem "know" which way is up? Do the roots "know" they should grow downward?

In this experiment, you'll discover whether seeds are able to respond to the forces of gravity (**gravitropism**).

 Note: This experiment requires at least a week to complete.

INQUIRY AND ANALYSIS

1. **Dampen** a paper towel, fold it in half, and wrap it around the inside of a clear, plastic cup. Press the paper towel along the inside of the cup so that it adheres to the sides.

   Place **four dried beans** between the paper towel and the side of the cup. Space the beans equally around the diameter of the cup. Position **two of the seeds vertically** and the other **two horizontally** (at a 90° angle to the table).

   With a permanent waterproof marker, **draw an** arrow on the outside of each cup that **matches the direction** of seed orientation, as shown in **Figure 13-3, parts a and b.** Number the seeds 1–4.

a. Vertical Orientation

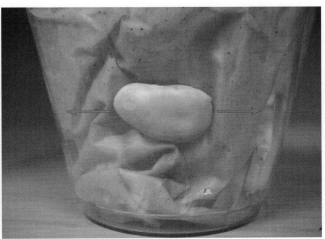

b. Horizontal Orientation

FIGURE 13-3 Setup for Gravitropism Experiment

2. Add **two tablespoons of tap water** to the bottom of the cup. Invert the second plastic cup on top (to serve as a lid and keep the setup moist). If the paper towel begins to dry out, add more water.

3. Check the experiment once a day and record your observations of growth in all four seeds.

   Record the initial **orientation** of each seed (**horizontal or vertical**).

   Seed 1 –

   Seed 2 –

   Seed 3 –

   Seed 4 –

4. Record your daily observations in **Tables 13-2a** and **13-2b.** Be sure to take note of the following, in addition to your other observations:

- sprouting sequence (root first, shoot first, leaves first, etc.)
- splitting of seed coat
- any changes in the growth direction of the root and shoot
- presence (and location) of root hairs
- color changes in the root, shoot, and leaves

TABLE 13-2a	Tropism Observations
Day 1	Seed 1: Seed 2: Seed 3: Seed 4:
Day 2	Seed 1: Seed 2: Seed 3: Seed 4:
Day 3	Seed 1: Seed 2: Seed 3: Seed 4:
Day 4	Seed 1: Seed 2: Seed 3: Seed 4:

	TABLE 13-2b	**Tropism Observations**
Day 5	Seed 1:	
	Seed 2:	
	Seed 3:	
	Seed 4:	
Day 6	Seed 1:	
	Seed 2:	
	Seed 3:	
	Seed 4:	
Day 7	Seed 1:	
	Seed 2:	
	Seed 3:	
	Seed 4:	
Day 8 (if needed)	Seed 1:	
	Seed 2:	
	Seed 3:	
	Seed 4:	

5. Based on the results of your experiment, do you agree or disagree with the following statement: "Gravitropism causes shoots and stems to grow up and roots to grow down no matter how the seeds are oriented." **Explain your answer.**

6. Based on the results of your experiment, what part of the bean plant exhibited **positive** gravitropism?

What part exhibited **negative** gravitropism?

7. What color were the outer leaves and stem when they first sprouted? _____

What color were they after several days? _____

What substance caused the color change? _____

After the color change occurs, what process can the plant carry out (that it couldn't do before the color change)?
_____

8. For the seed to first sprout, it was necessary to add _____ to each cup.

To support continued plant growth, you would also need:

Explain your answer.

ACTIVITY 3 • STRUCTURE AND FUNCTION

Almost all modern plants (more than 95%) are **angiosperms**—plants that produce flowers and contain seeds within a fruit. The basic organs of flowering plants are **stems, roots,** and **leaves.** The **shoot** extends above the ground and consists of the stem and leaves.

Energy storage

The leaves (and sometimes the stem itself) produce sugars through photosynthesis which are stored in the form of starch or converted into other organic molecules. Often, the starch is transported to other locations in the plant for storage.

INQUIRY AND ANALYSIS

1. Form a hypothesis about whether you expect starch to be present in a potato. Write the hypothesis below.

2. Describe a method you could use to test your hypothesis using supplies you've used in previous lab exercises.

3. Slice off a piece of potato and try it out! Was your hypothesis supported?

Surface area and absorption

The root system below ground provides support by anchoring the plant, absorbs water and dissolved minerals from the soil, and can serve as a location for nutrient storage. An examination of radish root hairs will demonstrate the structures present which aid in water and mineral absorption.

INQUIRY AND ANALYSIS

1. Play the section of the DVD entitled **"Radish Root Hairs."**

   How do root hairs help the root carry out its functions more efficiently?

2. If a plant were infected with a disease that destroyed its root hairs, what do you think might happen to the plant? Why? **Explain your answer.**

Biotechnology Today

A recent innovation in plant breeding uses plants that can absorb unusually large amounts of toxic substances without being harmed. These plants are used to clean up contaminated soil and water in a process known as **phytoremediation.** Toxic materials are picked up when contaminated water enters the root system. Absorbing high levels of toxic materials would kill most plants, but some species can detoxify or sequester the toxins up to 100 times more than the normal level without harmful effects.

Several species have demonstrated an amazing ability to absorb heavy metals (such as zinc, arsenic, or lead) without harm. Most of these are found in the *Brassica* (cabbage) family. Later, the plants are harvested and either incinerated or processed to recycle the metals. The procedure can be repeated as many times as necessary to bring the contamination down to allowable levels.

Some species can also degrade (break down) organic pollutants. During this process, chemical reactions occurring inside the plants convert absorbed pollutants into nontoxic materials, which can be released safely to the soil or water.

Transport of water and nutrients

A system of transport tubes extends throughout the stem and branches, carrying water and nutrients to all the parts of the plant. Leaves lose water to the environment through evaporation (a process called **transpiration**). Water is needed in the leaves for photosynthesis to occur. As water continues to evaporate, more water molecules are drawn upward into the leaves through straw-like structures called **xylem tubes.**

Nutrients produced by photosynthesis are transported downward to the stems and roots through **phloem tubes.** In this activity, you'll demonstrate the movement of water through the stem toward the leaves of a plant.

INQUIRY AND ANALYSIS

1. Add tap water to a tall glass so that it's half full. Place at least **20 drops of food coloring** (red or blue) into the glass and stir.

 Note: Continue adding food coloring until the water is intensely colored.

2. Select two healthy looking celery stalks **with leaves.**

Carefully cut **two centimeters** off the **bottom** of each celery stalk **while holding the end you're cutting under water. Immediately** place each stalk in the glass of colored water.

Record the time you place the celery stalks in the water: _____

3. Allow the celery to remain in the colored water for a **minimum of six hours or overnight.**

Observe the celery once **after 30 minutes** have expired and then **once an hour** for the remainder of the six-hour time period.

4. How long did it take for the leaves to change color? _____

5. At the end of the six hours (or overnight), cut another two centimeters off the bottom of the celery stalks. Observe the cut end of the stalk. Color **Figure 13-4** to show the distribution of the dye in the celery stalks.

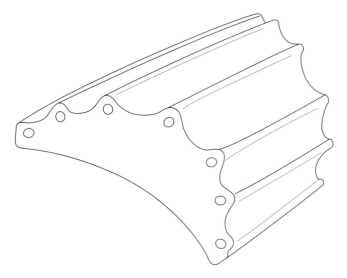

FIGURE 13-4 **Distribution of Dye in Celery Stalk**

6. Add a label for **xylem tubes** to your drawing in **Figure 13-4.**

7. How does the food coloring move up the stem into the leaves against gravity?

8. In early spring, trees convert stored starch from the stems into sugar. The xylem tubes transport this sugar, in addition to water, upward to where the leaf buds will form. Why is sugar transported in this direction at this time of year?

9. In early spring, stems that are damaged ooze a fluid called **sap.** The sap is collected from various types of maple trees and used to produce _____.

   (If needed, do a little Internet research to answer this question.)

10. As you've seen in your experiment, evaporation of water from the leaves draws additional water up through the stem (similar to how sweat cools your body). Transpiration and gas exchange in leaves occurs through small openings called **stomata.** When the temperature rises, the stomata open wider.

    Predict the effect of the following environmental conditions on **transpiration rate. Explain each answer.**

    - hot summer temperatures:

    - moist, humid air:

    - a windy day:

    - several days without rain:

ACTIVITY 4 • TREE RING ANALYSIS

Basics of tree ring analysis

The woody stems of trees continue to expand in diameter throughout the life of the tree. In temperate climates, most growth occurs during spring and summer with a dormant period during winter. This annual cycle can be seen as growth rings in a cross-section of the trunk. Tree rings form from the center of the tree outward. The ring **just inside the bark** of the tree is the **most recently formed.** The number of rings can be used to determine the age of the tree.

The rings form because of differences in cell size and cell thickness between the end of one year's growing season and the start of the next. When water is plentiful in spring, the wood has large diameter cells. This is called **early wood.** As moisture decreases during the summer, the cells are smaller and narrower (forming **late wood**). When looking at tree rings, the large diameter cells of the early wood appear light in color and the late wood looks darker, so one light ring plus one dark ring represents the annual growth in that tree (see **Figure 13-5**).

FIGURE 13-5 Annual Growth Rings in a Cross-Section of a Tree Trunk

INQUIRY AND ANALYSIS

1. **Find and highlight** the tree ring that represents the year the tree was planted.

2. If this tree was planted in 1953, when Watson and Crick discovered the structure of DNA, how many additional tree rings would be present? _____

3. **Find and highlight** the tree ring that represents the year:

    - of the first cell phone call   1973
    - of the Exxon Valdez oil spill   1989
    - that the first hybrid car hit the market 1999
    - of Hurricane Katrina   2005

4. If a winter had much more snow than normal, what effect would that probably have on the growth ring for the following spring? **Explain your answer.**

Analyzing core samples

To determine the age of a living tree, a small core of wood is removed from the trunk (see **Figure 13-6**).

FIGURE 13-6 Obtaining a Core Sample

Core sample patterns show the same light and dark rings you observed in the cross-section of the tree trunk (**Figure 13-7**). As with the tree rings in the cross-section, the rings in the core sample vary in width.

Wide rings are formed during years that had good growing conditions and narrow rings during years with poor growing conditions. For this reason, rings on core samples provide a historical record of the environmental conditions in a particular area—a glimpse into the past.

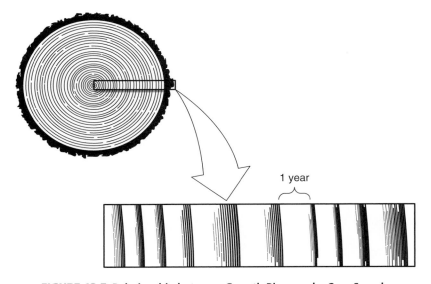

FIGURE 13-7 Relationship between Growth Rings and a Core Sample

INQUIRY AND ANALYSIS

1. The core sample illustration in **Figure 13-8** is a simplified version of a core taken from a living tree. Assume that the core sample was taken in **2010**. In which year was the tree planted? _____

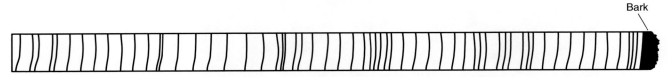

FIGURE 13-8 Core Sample

2. Color in the ring that represents the **year you were born.**

    Note: If the year you were born doesn't show in this core sample, color in the year Neil Armstrong first stepped onto the moon (1969).

   Was that a good growth year for the tree? **Explain your answer.**

3. Describe the growth pattern between **1979** and **1983**.

4. List three environmental factors that might have contributed to the growth ring pattern observed between **1979** and **1983**.

Exercise 14 • Homeostasis and The Circulatory System

OBJECTIVES

After completing this exercise, you should be able to:

- discuss the effect of external temperature changes on skin and body temperatures
- list and explain several homeostatic mechanisms that help to maintain physiological stability
- construct and interpret graphs that show the effects of exercise on the circulatory system
- explain the changes in pulse rate and body temperature that occur during and after exercise
- identify and explain the functions of each of the major structures of the heart
- trace the path of blood through the heart

SUPPLIES

Activity 1

SUPPLIES FROM LAB KIT

- strip thermometer, wide range, 1
- strip thermometer, narrow range, 1

HOUSEHOLD SUPPLIES

water, 2 gallons

basin or dish pan, 1

digital oral thermometer, 1

Activity 2

SUPPLIES FROM LAB KIT

- strip thermometer, narrow range, 1

HOUSEHOLD SUPPLIES

digital oral thermometer, 1

stop watch (or any watch or clock with a second hand), 1

Activity 3

HOUSEHOLD SUPPLIES

blue and red colored pencils (or highlighters), 1 each

Activity 4

- DVD: *Dissection of the Fetal Pig*

ACTIVITY 1 • UNDERSTANDING HOMEOSTASIS

The body's normal temperature is approximately 98.6°F (37°C). The body maintains this temperature within a narrow range, whether the outside temperature is hot or cold. The ability to do this is called **homeostasis.** Homeostasis literally means "steady state," indicating that the internal environment stays relatively the same regardless of changes in the external environment.

To illustrate how homeostasis works, you'll examine whether internal body temperature is affected by submerging your arm in water of different temperatures.

INQUIRY AND ANALYSIS

Read through the instructions completely before you begin the experiment.

1. Fill a basin or dish pan with **one gallon** of cold tap water. Place **20 ice cubes** in the basin. Wait **five minutes.** The temperature of the water in the basin will cool to 50°F. Measure the temperature with a **digital oral thermometer** and **record** your results in **Table 14-1.**

   Caution! For safety reasons, use only mercury-free thermometers.

   Note: The temperature range of most oral thermometers isn't low enough to record the temperature of the ice water. If your oral thermometer will work for this purpose, measure the temperature of the water in the basin and record your measurement in **Table 14-1.** If not, use the predetermined measurement of 50°F and record that number in the table.

2. Place the **narrow range strip thermometer** from your lab kit **halfway between your elbow and wrist on one of your forearms** and measure the temperature of the skin. Record your results in **Table 14-1.**

   Note: Take the temperature on the same arm that you plan to place in the cold water in step 3 below.

3. Place your forearm in the basin of cold water for **five minutes.**

   At the end of the five-minute period, **while keeping your arm in the basin of cold water,** use the oral thermometer to measure your body temperature. Record your results in **Table 14-1.**

4. Remove your arm from the water and blot it dry.

   Immediately, place the **wide range strip thermometer** from your lab kit **halfway between your elbow and wrist on the forearm that was in the cold water** and measure the temperature of the skin. Record your results in **Table 14-1.**

5. Using the **narrow range strip thermometer,** measure the temperature of the arm that wasn't in the basin of cold water. Measure the temperature **at the same place on your forearm** where you took the reading in step 3. Record your results in **Table 14-1.**

6. **Repeat steps 1 through 5,** with the following changes:

   Instead of cold water, use the **hottest** water from your faucet that you can **comfortably tolerate on your skin.**

   Caution! The water should not be hot enough to burn your skin.

   Measure the water temperature with the oral thermometer as before, but for the arm temperature measurements **use only the narrow range strip thermometer.** Record your temperature results in **Table 14-1.**

   TABLE 14-1 Temperature Results °F

Temperature Measurements	Cold Water Experiment	Hot Water Experiment
Water in basin		
Arm before test		
Core temperature (mouth)		
Submersed arm		
Dry arm		

7. Did the results of the cold and hot water experiment demonstrate homeostasis? **Explain your answer.**

8. Did this experiment include a **control**? If so, what was it? If not, why not?

9. Individuals suffering from anorexia experience a severe loss of body fat, which normally provides insulation to the body. These changes are often accompanied by an excess growth of body hair. Explain how this might be related to homeostasis.

ACTIVITY 2 • HOMEOSTASIS, EXERCISE, AND HEART RATE

Exercise can cause a variety of changes within the body. As you exercise, your muscles generate more heat. Mechanisms of homeostasis (such as **increased perspiration**) are activated to remove excess body heat and maintain body temperature.

Changes in blood flow also compensate for temperature changes during exercise. Blood vessels in the skin increase in diameter (**vasodilation**), which allows more blood flow to the skin and the removal of excess heat by diffusion. The reverse is also true. When exposed to cold temperatures, the blood vessels in the skin decrease in diameter (**vasoconstriction**), conserving heat in the body core.

Your heart rate (the number of beats per minute) can be determined by **taking your pulse** at specific regions of the body. One such location is the side of the neck, where the **carotid artery** is located.

In this activity, you'll investigate whether exercise causes changes in **body temperature** and **pulse rate**.

INQUIRY AND ANALYSIS

Read through the instructions completely before you begin the experiment.

1. Sit and relax for **one minute.** At the end of the minute, **while still sitting,** place the **narrow range strip thermometer** on your forehead. Record the temperature in **Table 14-2.**

   Remain sitting and use a **digital oral thermometer** to measure your body temperature. Record your results in **Table 14-2.**

    Caution! For safety reasons, use only mercury-free thermometers.

2. To find the pulse point on your carotid artery, take the index and middle fingers of either hand and place them at the end of your jaw, just below your ear. From this point, slowly draw your fingers down toward your neck until you feel a slight beating sensation against your fingers (see the correct location in **Figure 14-1**). This beat is your pulse.

Sit and relax for **one minute.** At the end of the minute, **while still sitting,** take your pulse for **one minute.** Record your results in **Table 14-2.**

FIGURE 14-1 Locating the Carotid Artery

3. Do jumping jacks for **four minutes.** This must be four minutes of **continuous** exercise (**no rest breaks**).

   Caution! If you have cardiovascular or breathing problems, ask someone else to do the exercise portion of this activity for you.

4. At the end of the exercise period, **immediately** take your pulse rate. Record your results in **Table 14-2.**

   As quickly as possible measure your **body and skin temperature** as outlined in step 1. Record your results in **Table 14-2.**

TABLE 14-2 Exercise Results

Exercise Condition	Skin Temperature	Oral Temperature	Pulse Rate
Rest			
Immediately after exercise			
Four minutes after exercise			
Eight minutes after exercise			

5. Record the same measurements **four minutes after** you stop exercising. Wait **another four minutes** and **record** the same parameters for a fourth time.

6. Compared to your resting temperature readings, were there any differences in your skin and oral temperature **immediately** after exercise? If so, give an explanation for any observed differences.

7. **Create a graph** showing the results of your experiment. Include all four readings for skin temperature, oral temperature, and pulse rate.

   Prepare the graph on a computer using the instructions in **Appendix I.** Submit the graph as required by your instructor.

   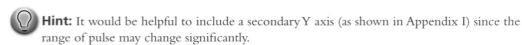 **Hint:** It would be helpful to include a secondary Y axis (as shown in Appendix I) since the range of pulse may change significantly.

8. In a few sentences, **summarize the changes** that occurred in these three parameters over the course of your experiment.

9. How is an increase in **pulse rate** related to homeostasis? In your answer, use the following terms: **oxygen, ATP production, cell respiration, blood flow,** and **heart rate**.

10. Predict the direction body temperature will change when the following conditions occur. Record your predictions in **Table 14-3.**

TABLE 14-3 **Body Temperature Predictions**

Condition	Predicted body temperature change (increase or decrease)
sweating	
vasoconstriction	
vasodilation	
shivering	

11. When you're participating in the following activities, predict which, if any, of the following conditions will occur:

    (Circle all conditions that apply.)

    Running to catch the bus on a hot summer day.
    sweating / vasoconstriction / vasodilation / shivering

    Sitting in a cold classroom listening to lecture.
    sweating / vasoconstriction / vasodilation / shivering

    Stepping out of a hot tub onto the patio in winter.
    sweating / vasoconstriction / vasodilation / shivering

    Sleeping under a heavy down quilt.
    sweating / vasoconstriction / vasodilation / shivering

12. On a cold winter's day, why do your ears, nose, fingers, and toes feel uncomfortably cold before your chest and abdomen? **Explain your answer.**

ACTIVITY 3 • STRUCTURES OF THE HEART

In the previous activity, you investigated the relationship between pulse rate and temperature as part of homeostasis. Pulse rate is directly related to activities that occur within the heart. The heart is the pump that circulates blood throughout the body. As the heart beats, blood is forced into the arteries and causes the arterial walls to expand. The expansion and relaxation of these walls is what we feel when we take our pulse.

In this activity, you'll examine the major structures of the heart and the vessels that bring blood to it (**veins**) and carry blood away from it (**arteries**).

Blood travels two full circuits through the heart. The **right side** of the heart sends blood to the **lungs.** This is referred to as the **pulmonary circuit.** The blood then returns to the **left side** of the heart from the lungs and is sent out to the **body tissues** (the **systemic circuit**). In both the pulmonary and systemic circuits, gas exchange occurs **only** in tiny blood vessels called **capillaries.**

When discussing the circulatory system, the color **blue** is used to indicate blood that is **low in oxygen and contains significant levels of carbon dioxide (deoxygenated).**

The color **red** is used to indicate blood that **has a significant level of oxygen and a lower level of carbon dioxide (oxygenated).**

INQUIRY AND ANALYSIS

1. In reference to **Figure 14-2,** color the **superior and inferior vena cava blue.**
   Why is the blood in these two vessels color coded blue?

    Hint: Where did this blood come from?

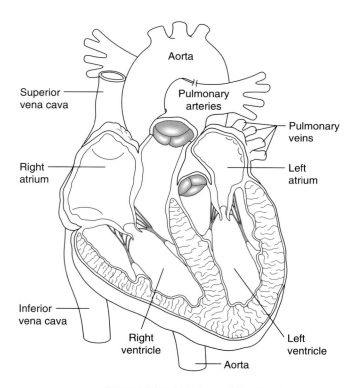

FIGURE 14-2 Structure of the Heart

2. Blood flows from each branch of the vena cava into the **right atrium.** Color this heart chamber the appropriate color (red or blue).

3. As the right atrium contracts, the blood is forced through the **tricuspid valve** and into the next chamber of the heart, the **right ventricle.** Valves between chambers of the heart prevent blood from flowing backward. Color this heart chamber the appropriate color (red or blue).

4. When the right ventricle contracts, blood will be forced through the **pulmonary semilunar valve** and into the **pulmonary arteries.** The pulmonary arteries will take blood **to the lungs** for gas exchange. Color the pulmonary arteries the appropriate color (red or blue).

5. Blood returning from the lungs through the pulmonary veins will enter the **left atrium.** Color this heart chamber the appropriate color (red or blue).

Hint: Remember that gas exchange occurred in the lungs.

6. As the left atrium contracts, blood is forced in through the **bicuspid (mitral) valve** and enters the **left ventricle.** Color this heart chamber the appropriate color (red or blue).

7. When the left ventricle contracts, blood will pass through the **aortic semilunar valve** and enter the **aorta** where it will be distributed to all parts of the body. Color the aorta the appropriate color (red or blue).

8. (Circle one answer.)

   The **right / left** side of the heart was colored blue.

   The **right / left** side of the heart has oxygenated blood.

9. In general, arteries carry oxygenated blood and veins carry deoxygenated blood. Which vessels on the diagram are **exceptions** to this rule? **Explain your answer.**

10. (Circle one answer.)

    The inferior vena cava is an **artery / vein.**

    The aorta is an **artery / vein.**

    Explain your answers.

11. There are two valves on each side of the heart. On each side, only one valve can be opened at a time. When one of the valves is opened, the other must be closed.

Which two valves are on the **right side** of the heart?

_____   _____

Which two valves are on the **left side of** the heart?

_____   _____

On the **right side** of the heart, which valve is located between **a heart chamber and a blood vessel?** _____

On the **left side** of the heart, which valve is located **between two heart chambers?**

_____   _____

12. When the right atrium contracts, the _____ valve will be

opened and the _____ valve will be closed.

When the left ventricle contracts, the _____ valve will be

opened and the _____ valve will be closed.

When blood travels to the lungs, the _____ valve will be opened.

When blood travels to the body tissues, the _____ valve will be opened.

13. Which side of the heart has a **thicker** muscular wall? How is this thicker wall related to the **function** of that specific side of the heart?

Biotechnology Today

New computerized scanning methods can provide early diagnosis and predict the probability of a patient developing coronary artery disease or having a heart attack. One such method is Electron-Beam Computed Tomography (EBCT), a safe, non-invasive technique to locate blockage in blood vessels attached to the heart. EBCT is a type of X-ray, but the scanner is much faster and more accurate than a traditional X-ray. Speed is critical when scanning the heart because it's in constant motion.

One application of EBCT technology is to measure calcium deposits in the coronary arteries. High levels of calcium deposits are known to be related to coronary atherosclerosis. The EBCT scan produces a coronary calcium score. A high score means a higher risk of heart problems over the next two years. A low calcium score, on the other hand, means that your risk of developing heart problems in the near future is low.

ACTIVITY 4 • CIRCULATION THROUGH THE FROG FOOT

As blood travels away from the heart, it passes through a series of smaller and smaller blood vessels, until it reaches the capillaries.

The diameter of capillaries is only slightly larger than the diameter of red blood cells themselves. **As blood vessel diameter decreases, the rate of blood flow decreases also.**

INQUIRY AND ANALYSIS

1. Play the section of the DVD, entitled **"Circulation through the Frog Foot."** In the video, you can see that the flow rate in capillaries is significantly slower than the rate of blood flow in the larger vessels nearby.

   (Circle one answer.)

   Circulation of blood through the frog foot is part of the **pulmonary / systemic** circuit.

   Circulation of blood through the capillaries is **faster / slower** than circulation through the larger vessels.

2. Complete the following sentence: A slow rate of flow is actually beneficial to the body because:

3. Several vessels in the DVD show the blood flow **slowing down and then speeding up.** What is causing the **change in flow rate?**

    Hint: Consider your experimental results from **Activity 2**.

Exercise 15 • Studying Organ Systems Through Dissection I

OBJECTIVES

After completing this exercise, you should be able to:

- apply your knowledge of anatomical terms to locate tissues and organs in the fetal pig
- distinguish between male and female pigs based on external features
- identify and explain the functions of each of the major organs and structures found in the mouth, throat, and neck
- trace the correct pathways for the movement of air and food through the mouth and throat
- explain how specialized features of neck organs assist in the functions of those organs

SUPPLIES

Activities 1–5

- DVD: *Dissection of the Fetal Pig*

ACTIVITY 1 • INTRODUCTION TO THE DISSECTION OF A FETAL PIG

The information gained from a close look at body organs can help you understand how body systems work and even some medical issues that might affect you or a family member in the future. A distance learning environment, however, makes it more difficult for you to experience dissection first-hand. You can come close to this experience by participating in a "virtual" dissection, using the accompanying DVD.

Pigs have a gestation period of about 16–17 weeks. At birth, fetal pigs range from 12–14 inches (30–36 cm) in length. The pigs used in the virtual dissection were approximately 9–11 (23–28 cm) inches in length.

INQUIRY AND ANALYSIS

1. Play the first section of the DVD, entitled **"Preparing for Your Dissection of the Fetal Pig."** Take note of the methods used to preserve the pigs and prepare them for use by students.

   The **umbilical cord** connects the fetal pig to the placenta. Compare the structure of the umbilical arteries and veins.

   (Circle one answer.) Arteries / veins have thick, muscular walls.

2. The fetal pig umbilical cord contains _____ arteries and _____ vein(s).

3. What substance(s) pass in through the umbilical cord to the fetus?

What substance(s) pass out through the umbilical cord to the mother?

ACTIVITY 2 • ANATOMICAL TERMS USED IN DISSECTION

Anatomical terms provide a basis for reference for describing the relationship of one body part to another. The terms used for four-legged animals are different from those used for humans, who walk upright.

The following are the basic terms used to describe the relative positions of body parts when dissecting a four-legged animal such as the fetal pig.

anterior	refers to the **head** end of the animal
posterior	refers to the **tail** end of the animal
dorsal	refers to the **back** of the animal
ventral	refers to the **belly** side of the animal

INQUIRY AND ANALYSIS

1. Play the section of the DVD entitled **"Anatomical Terms Used in Dissection."**

2. Refer to **Figure 15-1** and answer the questions below. In the DVD, you saw that the chin whiskers are anterior to the front legs. You also saw that the statement about relative positions can be reversed and it would also be correct to say that the front legs are posterior to the chin whiskers.

Using this same logic, the ears are _____ in relation to the tail.

The tail is _____ in relation to the ears.

The umbilical cord is _____ in relation to the spine.

The spine is _____ in relation to the umbilical cord.

The ear is _____ in comparison to mouth.

The mouth is _____ in comparison to the nose.

The pig on the tray is lying on its _____ side.

The tail is _____ in comparison to the umbilical cord.

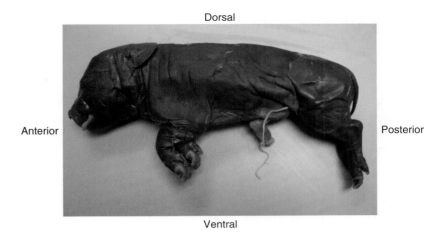

FIGURE 15-1 Anatomical Positions in the Fetal Pig

ACTIVITY 3 • EXTERNAL FEATURES AND GENDER DETERMINATION

In this activity you'll identify the external features of male and female pigs, and relate them to comparable human structures.

INQUIRY AND ANALYSIS

1. Play the section of the DVD entitled **"External Features and Gender Determination."** Look at the section that describes the external features of the **male pig.**

2. Take note of the location of the **scrotum** and **urogenital opening** in the **male pig.** Although both male and female pigs have a urogenital opening, they are found in different locations on the body.

   The two body systems referred to by the term "urogenital" are the _____

   system and the _____ system.

3. Continue to the section of the DVD that describes the external features of the **female pig.**

   Take note of the position of the **urogenital opening** (as compared to the location in male pigs).

   Do human females also have a common urogenital opening? _____

4. Locate the **labia** and **genital papilla** which are only found in females.

   Is it possible to distinguish between male and female pigs by the presence of **nipples**? **Explain your answer.**

ACTIVITY 4 • STRUCTURES OF THE MOUTH

Materials pass through the digestive and respiratory systems through openings located in the head. The relationships between these two important body systems can be studied by looking at structures from both systems that are present in the mouth.

 Note: Many body organs and structures have "special features" that enable them to perform their functions more effectively. As you proceed with the dissection, take some time to consider the relationship between each structure and how it functions in the body.

INQUIRY AND ANALYSIS

1. Play the section of the DVD entitled **"Structures of the Mouth."**

    Pause the DVD on the image of the **hard palate.** While looking at the screen, place your tongue on the roof of your mouth. Your tongue is now touching the hard palate.

    If the hard palate wasn't present, what body structures or organs might be vulnerable to injury from sharp objects in your mouth?

    Press **play** to complete this section of the DVD program.

    The **soft palate,** which is posterior to the hard palate, directs food toward the opening of the esophagus.

    As you know, the body is divided into right and left halves. The dividing line, called the **body midline,** can be seen running along the hard and soft palates. Sometimes this separation fails to fuse before birth. The resulting condition is known as a cleft palate.

2. Why is it important for cleft palates to be repaired?

    The **pharynx** connects the **oral and nasal cavities.** Through this opening, drainage from the nose can enter the mouth. This is why post-nasal drip can be irritating to the back of the throat.

3. After you've finished watching the DVD dissection of the mouth and jaw, **pause the DVD** and label the following structures in **Figure 15-2: hard palate, soft palate, body midline, tongue, epiglottis, pharynx,** and **glottis.**

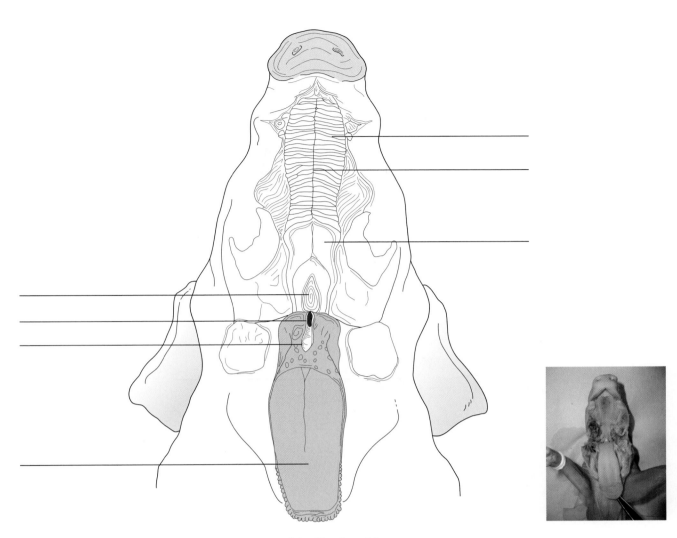

FIGURE 15-2 Structures of the Mouth and Jaw

The openings to the esophagus (which carries food to the stomach) and the trachea (which carries air to the lungs) are positioned close together at the back of the throat.

4. Refer to **Figure 15-3,** which shows the relative positions of the trachea and esophagus in the mouth and throat.

   Label the following structures on **Figure 15-3: nasal cavity, oral cavity, glottis, epiglottis, esophagus,** and **trachea.**

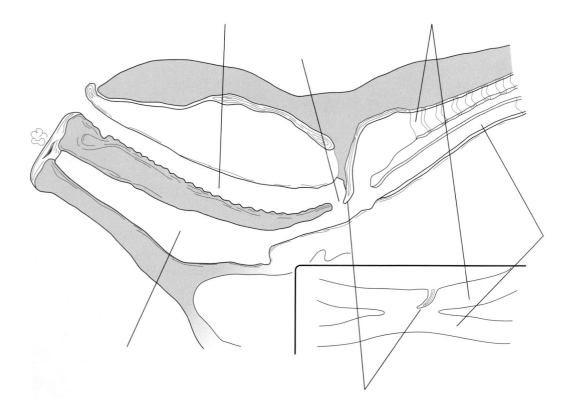

FIGURE 15-3 Pathways for Air and Food in the Fetal Pig

5. Using the diagram in **Figure 15-3, color the pathway of food** from the mouth to the esophagus.

   Using a different color, **trace the pathway of air** from the nasal cavity to the trachea. Notice that the **paths cross** each other.

6. **Draw an arrow** on **Figure 15-3** that shows the **direction** the **epiglottis has to move** to protect your airway when you're swallowing.

7. If you're talking and eating at the same time and begin to choke, what passageway did the food enter?

   What structure in the mouth prevents food and liquid from entering the air passageways?

   How does it work?

ACTIVITY 5 • ORGANS AND STRUCTURES OF THE NECK AND THROAT

The neck and throat contain organs and structures that represent a variety of body systems. The relative positions and structure of the organs of the pig throat are directly comparable to those found in humans.

INQUIRY AND ANALYSIS

1. Play the next section of the DVD, entitled **"Organs and Structure of the Throat."** Take note of the functions of each of the organs and structures you observe.

   Based on the information you gained from the DVD program, fill in the appropriate function or structure name in the blank spaces in **Table 15-1**.

   Answers can be used more than once.

TABLE 15-1 Organs and Structures of the Neck and Throat

Organ or Structure	Function
thymus	
	tube held open by cartilage rings
	this structure contains the vocal cords
thyroid gland	
vagus nerve	
	tube that transports food to the stomach
	tube that carries air to the lungs
	produces digestive enzymes in the mouth
carotid arteries	
jugular veins	

2. **Label** the following structures of the neck in **Figure 15-4**: **thymus, larynx, trachea, thyroid gland, carotid artery,** and **jugular veins.**

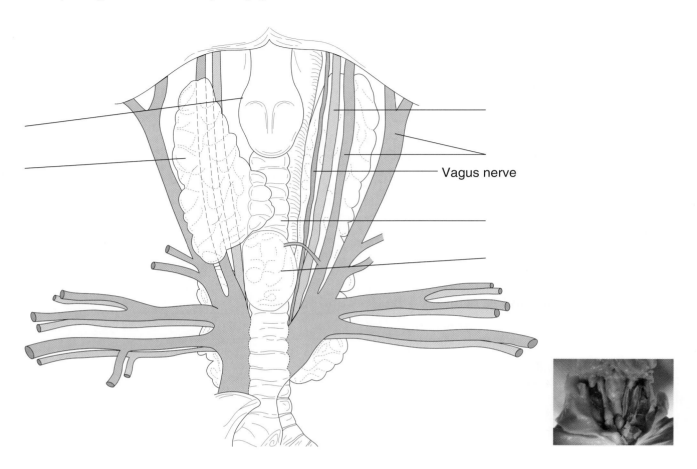

Vagus nerve

FIGURE 15-4 Organs of the Neck and Throat

3. List **all the body systems** that are represented by structure or organs of the neck or throat. Give **examples** for each body system.

4. Hold your chin tightly against your chest. Notice that you can't swallow food with your head in that position. However, you can breathe.

   Is the fact that you can **breathe, but not eat** when you lay down related to the differences in structure between the trachea and the esophagus? **Explain your answer.**

5. During mouth-to-mouth resuscitation the patient's chin is tilted up. How does this action help air get to the patient's lungs?

Biotechnology Today

Sci-Fi Surgery

You may have heard that surgeons are increasingly using medical robots to perform difficult surgical procedures. This can be especially useful when trying to access small or difficult-to-reach regions of the body. But did you know that the surgeons performing an operation can be thousands of miles away from the patient?

High-definition audio and visual signals travel underneath the ocean through fiber optic lines. Successful remote operations are now a reality. This technology opens the door for new applications, such as remote robot surgery on wounded soldiers in war zones or astronauts in space. It also means that patients around the world can have access to top surgical specialists through an Internet connection.

Exercise 16 • Studying Organ Systems Through Dissection II

OBJECTIVES

After completing this exercise, you should be able to:

- identify and explain the functions of each of the major organs and structures found in the thoracic cavity
- describe the mechanism by which breathing occurs
- identify and explain the functions of each of the major organs and structures found in the abdominal cavity
- explain how structural features of the stomach and small intestine increase the efficiency of function in those organs
- explain how malfunctions of various abdominal organs can affect your health

SUPPLIES

Activites 1–5

- DVD: *Dissection of the Fetal Pig*

ACTIVITY 1 • ORGANS OF THE THORACIC CAVITY

The two most prominent organs of the thoracic cavity are the lungs and heart, important parts of the respiratory and circulatory systems.

INQUIRY AND ANALYSIS

1. Play the section of the DVD entitled **"Organs of the Thoracic Cavity."** Take note of the position of each of the organs and structures you observe.

   Based on the information you gained from the DVD program, **label** the structures of the thoracic cavity in **Figure 16-1.**

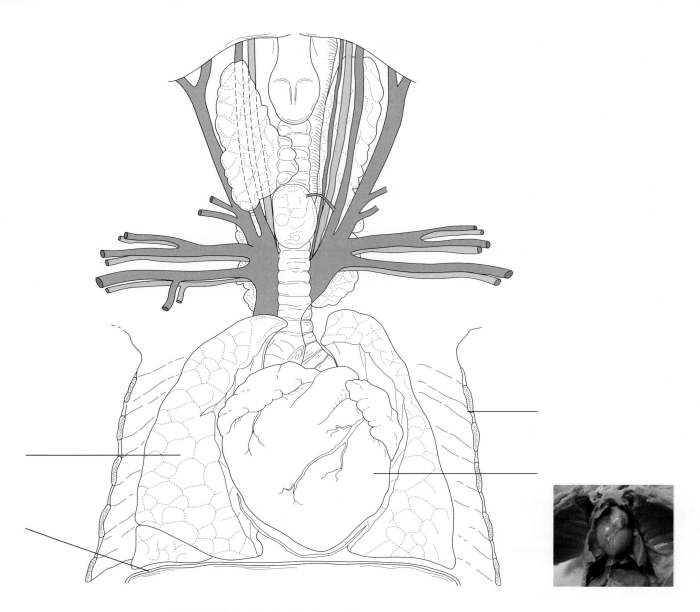

FIGURE 16-1 Organs of the Thoracic Cavity

You've just labeled the thoracic organs in Figure 16-1. This information can be put to practical use. When someone's choking, you can help them with the Heimlich Maneuver (see **Figure 16-2**). Here's how it's done:

- Make a fist and place the thumb side of your fist against the person's upper abdomen, below the ribcage and above the navel.
- Grasp your fist with your other hand and press it into their upper abdomen with a quick upward thrust.
- Alternatively, you can help yourself by leaning over a table edge, chair, or railing and pressing your upper abdomen against the edge to produce a quick upward thrust.

FIGURE 16-2 The Heimlich Maneuver

2. Based on the structure of the thoracic cavity in Figure 16-1 and the photo of the Heimlich Maneuver in Figure 16-2, why does the Heimlich Maneuver help solve a choking problem?

 Hint: Consider what you learned about the structures of the neck and throat in Exercise 15.

3. Advanced pregnancy or extreme obesity can prevent the diaphragm from completely flattening out. Would this affect your ability to breathe easily? **Explain your answer.**

ACTIVITY 2 • ABDOMINAL CAVITY

The organs of the abdominal cavity are separated from those of the thoracic cavity by the diaphragm. The majority of the body's organs are located in the abdominal cavity, and they carry out a wide range of functions.

INQUIRY AND ANALYSIS

1. Play the section of the DVD entitled **"The Abdominal Cavity."** Take note of the positions and functions of each of the organs and structures you observe.

Match each of the following organs or structures with the appropriate functions below. Answers can be used **MORE THAN ONCE**.

- **A.** spleen
- **B.** gall bladder
- **C.** liver
- **D.** stomach
- **E.** mesenteries
- **F.** large intestine
- **G.** small intestine
- **H.** bile duct
- **I.** rectum
- **J.** pancreas
- **K.** kidney
- **L.** urinary bladder
- **M.** peritoneum

____ storage of energy reserves

____ produces hormones that control blood glucose levels

____ removes damaged red blood cells from circulation

____ most chemical digestion occurs here

____ stores feces for elimination

____ stores bile

____ produces bile

____ support structure for blood vessels in the digestive system

____ bile passes through this structure when you eat foods containing fat

____ organ located between the esophagus and the small intestine

____ stores urine for elimination

2. If a person had their gall bladder removed, why might the doctor advise them to switch to a low-fat diet?

3. If you sustained an injury to your pancreas, how might this affect your ability to:

   a. digest foods:

   b. maintain homeostasis:

ACTIVITY 3 • THE STOMACH

The stomach is a small organ, but it has the ability to stretch and accommodate a large meal. In this activity, you'll examine the various specialized features present in this organ.

INQUIRY AND ANALYSIS

1. Play the section of the DVD entitled **"The Stomach."** Based on what you learned from the DVD program, answer the following questions.

   (Circle one answer.)

   The **cardiac sphincter / pyloric sphincter** is located between the stomach and the esophagus.

   The **cardiac sphincter / pyloric sphincter** is larger and stronger.

   When food enters the small intestine, the **cardiac sphincter / pyloric sphincter** is open.

   When you have heartburn, the **cardiac sphincter / pyloric sphincter** has opened.

2. You visit your family for Thanksgiving and eat an enormous amount of food. How does the **shape** of your stomach change to allow this? Explain the anatomical features involved.

ACTIVITY 4 • THE SMALL INTESTINE

The small intestine has an important role in both digestion and absorption of nutrients within the body.

INQUIRY AND ANALYSIS

1. Play the section of the DVD entitled **"The Small Intestine."** Take note of the various specialized features present in this organ.

    The evening news had a report about a boy who was born with a rare developmental abnormality—no villi in the small intestine. How would that affect his ability to absorb nutrients from his food? **Explain your answer.**

2. Gastric bypass surgery is performed on individuals that are seriously overweight. In this surgery, the stomach is made smaller and food bypasses a portion of the small intestine. **List and explain two reasons** why individuals who have this type of surgery should lose weight.

    a.

    b.

Biotechnology Today

There's a new technology that gives an entirely new meaning to "blue light special." Narrow band imaging (NBI) in conjunction with endoscopy is helping to see the interior of organs in much greater detail.

For example, a narrow band light source (mainly blue light) is attached to an endoscope and used to examine the GI tract. An endoscope consists of a small camera and light attached to a long, thin tube. The endoscope allows doctors to see inside the body without the need for surgery. When inserted through the mouth, the endoscope can examine the esophagus, stomach, small intestine, and part of the large intestine. The blue wavelengths are especially effective at identifying possible cancers that form in the esophagus of patients with advanced cases of acid reflux disease.

ACTIVITY 5 • THE URINARY SYSTEM

The urinary system not only helps remove metabolic wastes from the body, it also plays an important role in maintenance of ion balance and homeostasis.

INQUIRY AND ANALYSIS

1. Play the section of the DVD entitled **"The Urinary System."** Take note of the positions and functions of each of the organs and structures you observe.

   Based on the information you gained from the DVD program, fill in the appropriate function or structure name in the blank spaces in **Table 16-1**.

TABLE 16-1 **The Urinary System**

Organ or Structure	Function
	membrane that holds the kidneys in position
ureter	
	central region of the kidney that collects urine
cortex	
	stores urine before elimination
urethra	
urogenital opening	

2. **Label** the following structures on **Figure 16-3**: **kidney, medulla, cortex, ureter, urethra, urogenital opening,** and **urinary bladder.**

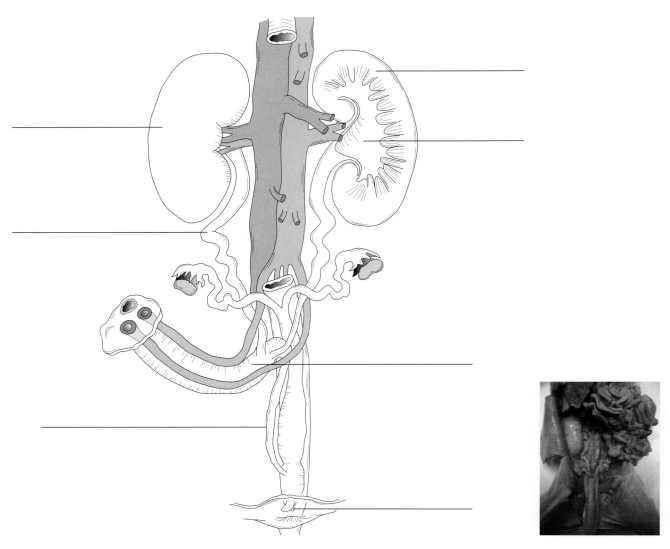

FIGURE 16-3 Urinary System of the Female Pig

3. Over the course of the day, the bladder fills with urine. What type of structure keeps the urine contained in the bladder?

4. People that suffer kidney failure need dialysis treatments. What function of the kidney is the dialysis machine replacing?

Exercise 17 • Metabolism and Nutrition

OBJECTIVES

After completing this exercise, you should be able to:

- explain basal metabolic rate (BMR) and its relationship to your daily caloric expenditures
- explain how diet and activity level help determine how effectively you lose weight and maintain your weight loss
- select foods based on their nutrient content and food label information
- suggest ways to improve your diet through alternative, healthier food choices

ACTIVITY 1 • CALORIES IN, CALORIES OUT

There have been several highly publicized diet programs (such as the low-carb diet) that claim to provide significant, permanent weight loss. In most cases, some weight loss occurs, but isn't sustained over a long period of time. It seems clear that basic changes in the typical American diet (for example, Figure 17-1) are needed to counteract the current "epidemic" of obesity and related health problems. Even young children have been affected.

FIGURE 17-1 Fast Food Meal

To lose weight, you must consume fewer calories than your body uses for metabolism and daily activities. **Calories** are a measurement of energy. Because one calorie is a very small unit, food calories are usually measured in units of 1,000 calories, called **kilocalories** (abbreviated **kcal**). The "magic formula" for weight loss is a simple equation. **One pound of fat** represents **3,500 calories** that the body has stored for later use. To lose a pound of fat, therefore, 3,500 kcal must be expended. The rate of energy consumption by the body is referred to as **metabolic rate,** which has two components:

- Your metabolic rate consists first of those calories used for basic body maintenance processes such as cell respiration, blood circulation, and removal of wastes. The total number of calories used for basic maintenance is called your **basal metabolic rate (BMR).**
- Although basal metabolic rate includes the majority of calories expended each day, it does **NOT** include calories used for **daily activities** (walking, studying, shopping, taking notes, etc.). This is the second component of your daily energy consumption.

In this activity, you'll be calculating your approximate energy expenditure for a typical day and comparing it with your typical daily caloric intake.

INQUIRY AND ANALYSIS

1. Select a **typical day** when you eat your normal number of meals. **Record** everything that you eat (including amounts).

2. Based on the information at the following web site, calculate the approximate number of calories consumed for all the food you recorded on your chart.

    www.mypyramidtracker.gov

    My **total caloric intake** over the recorded 24-hour period was _____ kcal.

3. Compare your calorie intake, as calculated above, to your calorie expenditures for BMR and daily activities. To calculate your BMR, use the appropriate formula for your gender.

Step 1 - Basic BMR
Female:

BMR = 655 + (4.354 × weight in lbs) + (4.569 × height in inches) − (4.7 × age in years)

Male:

BMR = 66 + (6.213 × weight in lbs) + (12.69 × height in inches) − (6.8 × age in years)

My BMR = _____ kcal.

Step 2 - The Activity Multiplier
Even though activity calories aren't part of your BMR caloric expenditures, your BMR calculation will be more accurate if it's modified for your general activity level because increased physical activity raises your overall metabolic rate (see an example in **Figure 17-2**).

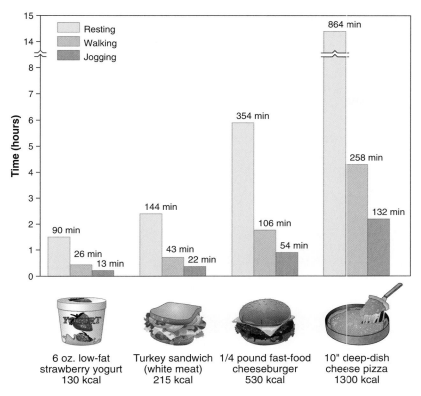

FIGURE 17-2 Burning Food Energy with Exercise

Your total caloric expenditure equals the calories expended for basal metabolism (BMR) plus the calories expended for daily activities. You can get a close estimate of the calories expended on daily activities by multiplying your BMR by the appropriate activity factor from the following list.

- sedentary (desk job, with little or no exercise) = BMR × 1.2
- lightly active (light exercise, 1–3 days/week) = BMR × 1.4
- moderately active (moderate exercise, 3–5 days/week) = BMR × 1.6
- very active (intensive exercise, 6–7 days/week) = BMR × 1.7

My **total caloric expenditure** (BMR times the selected activity multiplier) = _____ kcal.

4. Calculate your **energy balance** as follows:

total kcal consumed − total kcal expended = _____ kcal

5. **(Circle one answer.)**

My answer to question 4 above was a **positive / negative** number.

Therefore, I consumed **fewer / more** calories than I expended.

If this day's diet and activity are typical, over time I will probably **lose weight / gain weight / remain the same weight.**

6. If you consumed **100 kcal more than you expended** each day, how many days would it take you to accumulate 3,500 excess calories (and gain one pound)?

7. Return to your energy expenditure calculations and recalculate what your total calorie expenditures would be if you increased your activity multiplier by one level (for example, from light to moderate activity).

    If you were already at the highest activity level, recalculate for one level lower.

    How does the new figure change your energy balance equation?

8. If your goal was to **gain weight,** what changes could you make in your daily diet to improve your energy balance situation?

9. If your goal was to **lose weight,** what changes could you make in your daily diet to improve your energy balance situation?

10. Based on **calorie considerations alone,** which dieting strategy should be more effective for weight loss: a low carb diet or a low fat diet? **Explain your answer.**

    Hint: Carbohydrates and proteins contain 4 kcal/g. Fats and oils contain 9 kcal/g.

11. Would changing from a low carb to a low protein diet have any weight-loss effect? **Explain your answer.**

12. Based on your calculations, explain why athletes often gain a lot of weight when they retire from sports.

ACTIVITY 2 • GROCERY STORE SCAVENGER HUNT

INQUIRY AND ANALYSIS

1. Visit a local grocery store and find the items listed in **Table 17-1.** For each food you record in the table, include the **product name and a description.**

   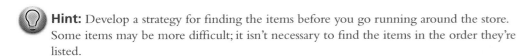**Hint:** Develop a strategy for finding the items before you go running around the store. Some items may be more difficult; it isn't necessary to find the items in the order they're listed.

2. For **food #8** in **Table 17-1,** explain why your food is "heart healthy."

3. For **food #13** in **Table 17-1,** explain why this food would contribute to high blood pressure.

4. For **foods #4, 5,** and **15** in **Table 17-1,** suggest **substitutes** for these foods that would be part of a **healthier** diet.

TABLE 17-1 Grocery Store Scavenger Hunt Results

Food Characteristics	Product Information
1. **Low fat, vegan** (no animal foods), **protein** source	
2. **Dairy** food suitable for a person who is **lactose-deficient**	
3. Food **high in fiber** (>10% daily value) and **low in fat** (<5% daily value)	
4. Packaged, **non-meat** food containing **saturated** fat	
5. Packaged food containing **trans fat**	
6. Food containing **lipids,** which serve as an **energy source** for a developing **embryo**	
7. Food you would avoid if you were watching your **cholesterol** intake	
8. "**Heart-healthy**" snack that **doesn't** require refrigeration, that you could pack in the morning to eat in between classes	
9. Food containing a **polysaccharide** that can be used as energy for **cell respiration**	
10. **Salad dressing** with **<1.5 grams of fat** per serving	
11. Food containing starch produced by **photosynthesis**	
12. Food that contains omega-3 fatty acids	
13. Food you would avoid if you were trying to control your blood pressure	
14. Food that contains a **mineral** needed to manufacture the protein hemoglobin	
15. Food that contains **tropical oils**	
16. Food containing high levels of **cellulose**	
17. Food that's produced by using **microorganisms**	

Exercise 18 • The Nervous System

OBJECTIVES

After completing this exercise, you should be able to:

- describe the functions of each of the following: sensory neuron, motor neuron, and interneuron
- list the steps of the spinal reflex arc
- explain why a reflex response is faster than a reaction time response
- analyze and present reaction time data
- predict how physiological factors can cause changes in the reaction response, and
- explain why afterimages appear in complimentary colors

SUPPLIES

Activity 2

SUPPLIES FROM LAB KIT
- None needed

HOUSEHOLD SUPPLIES
- a partner

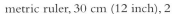

metric ruler, 30 cm (12 inch), 2

plastic wrap, 1 roll

cotton balls, 12

Activity 3

SUPPLIES FROM LAB KIT
- None needed

HOUSEHOLD SUPPLIES
- a partner
- chair, straight-back

metric ruler, 30 cm (12 inch), 1

Activity 4

SUPPLIES FROM LAB KIT
- None needed

HOUSEHOLD SUPPLIES
- a partner

bright blue paper, 1 sheet

white paper, 1 sheet

bright yellow paper, 1 sheet

stopwatch (or any watch or clock with a second hand)

ACTIVITY 1 • STEPS OF THE SPINAL REFLEX

The nervous system is made up of two parts:

- the **central nervous system,** which consists of the brain and spinal cord
- the **peripheral nervous system,** which includes all the remaining nervous system structures

Nerve cells are called **neurons.** Neurons can have several different functions. **Sensory neurons** receive incoming information about the environment and transmit it to the central nervous system (the brain and spinal cord).

Motor neurons carry instructions from the central nervous system to muscles and glands.

Interneurons, which are located only within the brain and spinal cord, analyze sensory signals and relay instructions to motor neurons.

Reflexes are **involuntary motor responses.** A familiar spinal reflex occurs when you accidentally step on a sharp object. Before you even realize that your foot hurts, a reflex has lifted your foot off the sharp point. During a **spinal reflex**, a sensory receptor is stimulated and the impulse is transmitted to the spinal cord (see **Figure 18-1**).

Most spinal reflexes are controlled directly by the spinal cord. The spinal cord triggers motor neuron responses without input from the brain, although the brain can influence the strength of reflex responses. The signal continues on to the brain from the spinal cord, but by the time you're consciously aware a problem exists, the **corrective action has already been triggered** by the spinal cord.

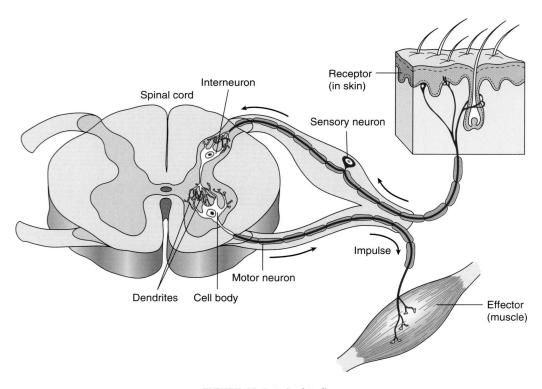

FIGURE 18-1 Spinal Reflex

INQUIRY AND ANALYSIS

1. You're preparing dinner at home, when you accidentally touch a hot pan. The following activities will be occurring in your nervous system. Place the steps of the reflex and response actions in their correct order by **numbering them 1–9.**

   _____ muscles in your arm and hand contract

   _____ sensory neurons are stimulated

   _____ brain senses pain

   _____ sensory neurons transmit impulse to spinal cord

   _____ motor neurons transmit impulses to hand and arm muscles

   _____ reflex action—hand moves away from pan

   _____ hand touches hot pan

   _____ you go over to the sink to run your hand under cold water

   _____ you put a bandage on the burn

Biotechnology Today

Communication is possible even in organisms that don't have a nervous system. Neurons have receptors so they can receive chemical signals from other similar cells. An amazing recent discovery is that bacteria also communicate by releasing small signaling molecules. Bacteria use their communication ability to "talk" with one another and work together in groups.

Bacteria form communities called biofilms. You experience a biofilm every morning when you wake up and your teeth are covered with a "fur coat." That coating contains over 600 species of bacteria, arranged into specific layers. Even though you brush your teeth and remove the coating, it forms again in exactly the same arrangement the next day.

What use is the ability to communicate to all different types of bacteria? They change their behavior depending on population density and can work together as a community, just as the cells of multicellular organisms can do. And in both cases, cells working together can be more effective and improve the chances of survival for the individual.

ACTIVITY 2 • THE BLINK REFLEX

Muscles attached to the eyelids allow blinking of the eyes. The "blink reflex," which occurs automatically, helps to protect the eyes from foreign objects. To demonstrate the blink reflex, perform the following experiment.

INQUIRY AND ANALYSIS

1. Get the following supplies: **clear plastic wrap and several cotton balls.**

   Crush the cotton together into one large ball, about the size of a golf ball.

2. **Work with a partner** to complete this experiment. You'll be the test subject and your partner will perform the experiment and record the results.

   Wrap each end of the plastic wrap twice around a ruler. The distance between the two rulers should be about **12 inches (30 cm).**

   Pull the rulers apart so that the plastic wrap is stretched tightly between them. Hold the rulers and plastic wrap about **six inches (15 cm)** in front of your face and eyes (as shown in **Figure 18-2**).

FIGURE 18-2 Blink Reflex Setup

3. Relax, but remain alert. Your partner will toss the cotton ball at the plastic wrap and observe your blink response (this is **Trial 1**). **Record your results in Table 18-1.**

   Repeat the procedures until you've completed a **total of ten trials. Record the results of each trial in Table 18-1.**

4. **Exchange roles** and repeat steps two and three.

TABLE 18-1	Results of Blink Experiment	
Trials	Blink (Yes/No) Subject 1	Blink (Yes/No) Subject 2
1		
2		
3		
4		
5		
6		
7		
8		
9		
10		
Total (number of blinks)		

5. Using the following formula, calculate the percentage of blink responses for each test subject.

$$\frac{\text{total number of blinks}}{\text{total number of trials}} \times 100 = \underline{\hspace{2cm}} \%$$

Subject 1 blink responses = _____ %

Subject 2 blink responses = _____ %

6. Were there any differences in the blink response between you and your partner? If so, **explain** how each of you responded differently.

7. Since you knew that the plastic wrap would prevent the cotton ball from hitting your face, why did the blink response occur? **Explain your answer.**

8. List **two** situations in everyday life when the blink reflex would be activated in humans or pets. **Explain your answer.**

ACTIVITY 3 • REACTION TIME

A reaction is a **conscious, voluntary response** to an outside event. You exert conscious control of your muscles when you chew food, manipulate your knife and fork, or take notes in class.

During a voluntary response, sensory receptors are stimulated and the impulse is transmitted through the spinal cord to higher brain centers. The incoming signal and response **travel a greater distance** than in a spinal reflex (which is mediated by the spinal cord), therefore, a reaction response **requires more time** to complete than a reflex.

The object of this experiment is to see how quickly you can catch a falling ruler (a measure of your **reaction time**).

Read through the instructions **completely** before beginning this experiment.

INQUIRY AND ANALYSIS

1. Work **with a partner.** You're the test subject. Sit on straight chair (such as a kitchen chair).

    Your partner is the investigator conducting the reaction time experiment. He or she will stand **facing you,** on the side of the chair that **matches your dominant hand** (right-handed or left-handed).

    The investigator will **hold the ruler vertically,** with the **one-centimeter end down** (toward the floor), about **one hand-width above your dominant hand** (see **Figure 18-3**). **The numbers on the ruler should be facing the investigator.**

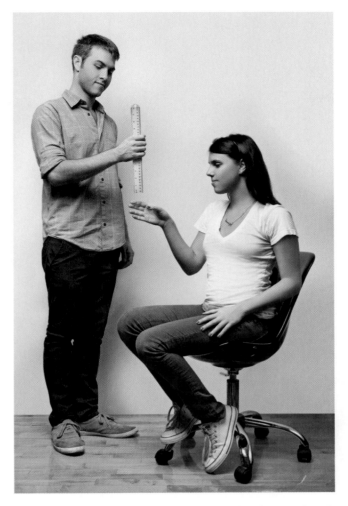

FIGURE 18-3 Position of the Investigator and Test Subject for the Reaction Time Experiment

2. You (the test subject) should position the **thumb and fingers of your dominant hand** about an **inch apart,** ready to catch the ruler (see the example in **Figure 18-4**.)

3. When you're ready to begin, say "ready." Within 10 seconds (you won't know **exactly** when), the investigator will drop the ruler and you'll try to **catch it.**

   After catching the ruler—keep holding it!

   The investigator will observe the **centimeter mark closest to where you caught the ruler** (rounded off to the nearest centimeter).

   Record the centimeter mark in **Table 18-2** in the column marked **Distance (cm).**

   The centimeter mark on the ruler can be **converted to reaction time** by using **Table 18-3.**

   Record the reaction time for each trial in **Table 18-2** in the column marked **Reaction Time (sec).**

ACTIVITY 3 • REACTION TIME

FIGURE 18-4 Hand Position for Reaction Time Experiment

TABLE 18-2	Results of Reaction Time Experiment			
	Subject 1		Subject 2	
Trial	Distance (cm)	Reaction Time (sec)	Distance (cm)	Reaction Time (sec)
1				
2				
3				
4				
5				
6				
7				
8				
9				
10				
Mean				

TABLE 18-3	Reaction Time Conversions				
Distance (cm)	Time (sec)	Distance (cm)	Time (sec)	Distance (cm)	Time (sec)
1	0.045	11	0.150	21	0.207
2	0.064	12	0.156	22	0.212
3	0.078	13	0.163	23	0.217
4	0.090	14	0.169	24	0.221
5	0.101	15	0.175	25	0.226
6	0.111	16	0.181	26	0.230
7	0.120	17	0.186	27	0.235
8	0.128	18	0.192	28	0.239
9	0.136	19	0.197	29	0.243
10	0.143	20	0.202	30	0.247

4. Repeat the procedure **nine more times**, recording the results of each trial in **Table 18-2**.

   Calculate the **mean distance and reaction times** for Test Subject 1. Enter the results in Table 18-2.

5. Switch places with your partner and **repeat** steps two through four.

6. Using your experimental results, **create a graph** that compares the **number of trials and the reaction times** for the two test subjects.

   Prepare the graph on a computer using the instructions in **Appendix I. Submit** the graph as required by your instructor.

7. **Calculate** the **mode** of your reaction times **(most frequently occurring number)** and determine the **range** of the data **(difference between highest and lowest numbers)**.

   mode = _____   range = _____

8. Suggest a method that could be used to determine whether your **average reaction time is faster or slower** than that of a typical college student.

9. How do you think your reaction time would change if you repeated the test at the following times of day? **Explain each answer.**

   a. 7:00 AM:

   b. 5:00 PM:

   c. 2:00 AM:

   d. after drinking a double black coffee or a can of Red Bull cola:

   e. after taking two antihistamine tablets (such as cold medications or Benedryl®):

10. You have a Monday-night football party and your guests have a few drinks. Would it be a good idea to administer the ruler reaction time test before you let someone drive their car home? **Explain your answer.**

ACTIVITY 4 • VISUAL PERCEPTION

The retina contains sensory receptors called **rods** and **cones**. Rods are more sensitive at low light levels, but human color vision relies on cones. There are three types of cones: **blue, green,** and **red.**

Each type of cone receives different wavelengths from the visual spectrum of light. Any visible color can be produced by different combinations of the **three primary optical colors:**

red, blue, and green (as received by different combinations of cones). **Equal stimulation** of all three types of cones is perceived as **white** light (see **Figure 18-5**).

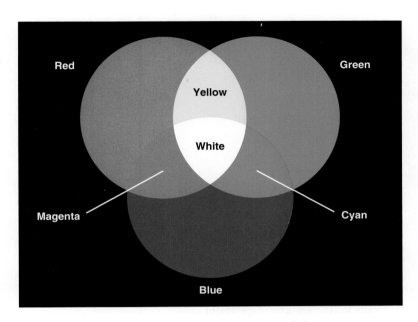

FIGURE 18-5 Stimulation of Red, Green, and Blue Cones

Complementary colors appear directly opposite each other on a color wheel. Looking at **Figure 18-5,** you can see that **red** and **green (or blue-green)** are opposite each other and, therefore, are complementary colors.

Continuous exposure to light of a particular wavelength causes the cones responding to that particular wavelength to become fatigued. Visual reception is shifted to the cones that receive the complementary color. For example, if you stare at a red spot, the red-sensitive cones will become fatigued. The "red" signal to the brain gradually decreases. So when you shift your gaze to a white surface, the image appears green. The retention of a previous image after the source has been removed is called an **afterimage**.

To demonstrate this phenomenon, try an experiment with two complementary colors, blue and yellow.

INQUIRY AND ANALYSIS

Read through the instructions **completely** before beginning this experiment.

1. **Work with a partner.**

    Cut a square of **bright blue** paper approximately **three inches by three inches** (7.5 × 7.5 cm) in size.

    Lay a sheet of white paper horizontally on a flat surface. Place the colored square towards one side of the sheet of white paper, leaving a border of white between the colored paper and the edge of the white paper (see **Figure 18-6**).

    Shine a **bright light** on the paper.

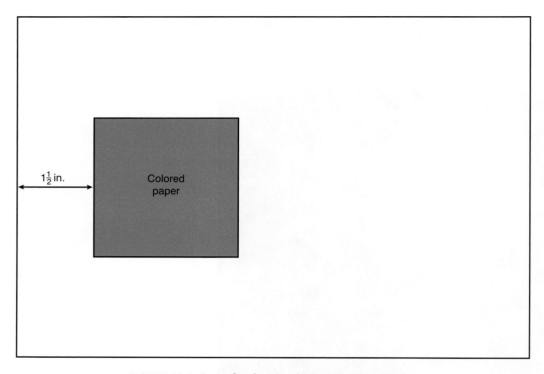

FIGURE 18-6 Setup for the Visual Perception Experiment

2. Stare directly at the **blue section** of the paper while your partner records your times for **60 seconds**.

At the end of the 60 seconds, shift your eyes and stare directly at the **white section** of the paper, while **your partner continues to time you.**

If you see an afterimage, **tell your partner** when it **disappears.**

Did you see an afterimage? _____ If so, what color was it? _____

Approximately how long did the afterimage last? _____

3. Repeat steps 1 and 2 using a square of **bright yellow** paper.

Did you see an afterimage? _____ If so, what color was it? _____

Approximately how long did the afterimage last? _____

4. You're taking photos at a friend's birthday party. Your friend complains that you're "blinding him with the flash" and when he closes his eyes, he can still see the flash. Is there a biological basis for his assertion? **Explain your answer.**

Exercise 19 • The Reproductive System

OBJECTIVES

After completing this exercise, you should be able to:

- identify the organs and structures of the male and female reproductive systems
- explain the functions of each organ and structure in the male and female reproductive systems
- explain the role of male accessory organs in the reproductive process
- analyze and interpret data collected in the simulated transmission of an infection

SUPPLIES

Activities 1-2

- DVD: *Dissection of the Fetal Pig*

Activity 3

SUPPLIES FROM LAB KIT

- None needed

HOUSEHOLD SUPPLIES

white and dark beans (or different colored candies, checkers, etc.), 48 each color

ACTIVITY 1 • THE MALE REPRODUCTIVE SYSTEM

Although the male and female reproductive systems appear very different externally, there are internal similarities. Both sexes have pairs of **gonads** (the testes in males and the ovaries in females). Within the gonads, the process of meiosis occurs, producing sex cells (sperm and eggs). In addition, both sexes have accessory organs that contribute to the reproductive process.

In this activity, you'll become familiar with the organs and structures of the male reproductive system.

INQUIRY AND ANALYSIS

1. Play the **first section** of **"The Reproductive System"** on your DVD, entitled **Male Anatomy.** Take note of the locations and functions of each of the organs and structures you observe.

   Based on the information you gained from the DVD program, **label** the following organs and structures of the male reproductive system in **Figure 19-1: testes, epididymis, vas deferens, scrotum, inguinal canal, seminal vesicles, urethra, urogenital opening,** and **bulbourethral (Cowper's) glands.**

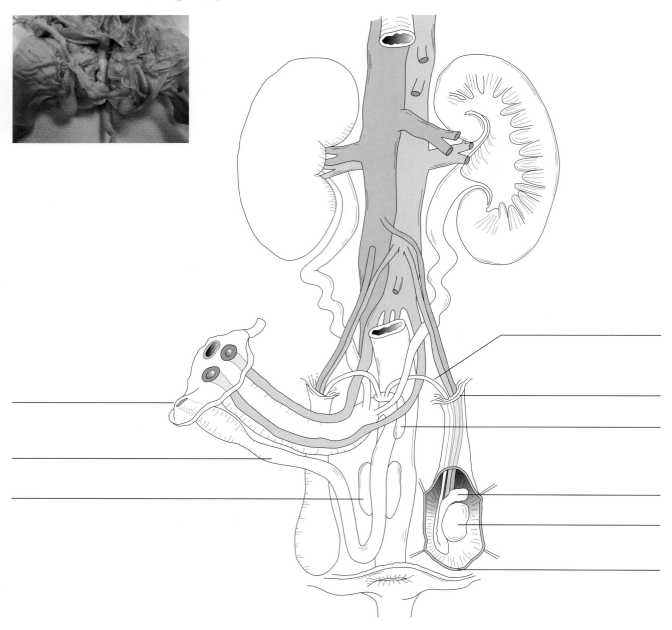

FIGURE 19-1 Reproductive System of the Male Pig

2. On **Figure 19-1,** write the **letter S** in the location where sperm are produced.

   On **Figure 19-1,** write the **letter E** in the location where sperm are stored before ejaculation.

   On **Figure 19-1,** write the **letter F** in two locations where fluid is produced for ejaculation.

3. What two substances travel through the urethra of a male pig?

   _____   _____

4. If the testes didn't descend into the scrotum, how would sperm production be affected? **Explain your answer.**

5. How would the ejaculated fluid of a male who has had a vasectomy (an operation that closes the vas deferens) differ from the fluid ejaculated by the same man before the operation?

6. As men age, the prostate may enlarge and make urination difficult. How can an enlarged prostate cause this problem?

ACTIVITY 2 • THE FEMALE REPRODUCTIVE SYSTEM

In this activity, you'll become familiar with the organs and structures of the female reproductive system.

INQUIRY AND ANALYSIS

1. Play the **second section** of the **"Reproductive System"** on the DVD, entitled **Female Anatomy.** Take note of the positions and functions of each of the organs and structures you observe.

    Based on the information you gained from the DVD episode, **label** the following organs and structures of the female reproductive system in **Figure 19-2: ovary, fallopian tube, horns of the uterus, body of the uterus, urogenital opening,** and **vagina.**

2. On **Figure 19-2,** write the **letter F** in the location where fertilization of eggs would occur.

    On **Figure 19-2,** write the **letter I** in the location where implantation of embryos would occur.

    On **Figure 19-2,** write the **letter S** in the location where sperm would be deposited by a male pig.

3. If you have your ovaries removed for medical reasons, hormone replacement therapy is usually needed. Would hormone replacement therapy also be required after a tubal ligation? **Explain your answer.**

4. On **Figure 19-2,** use a highlighter to trace the pathway of a sperm from its entry to the female reproductive tract to the location where fertilization occurs.

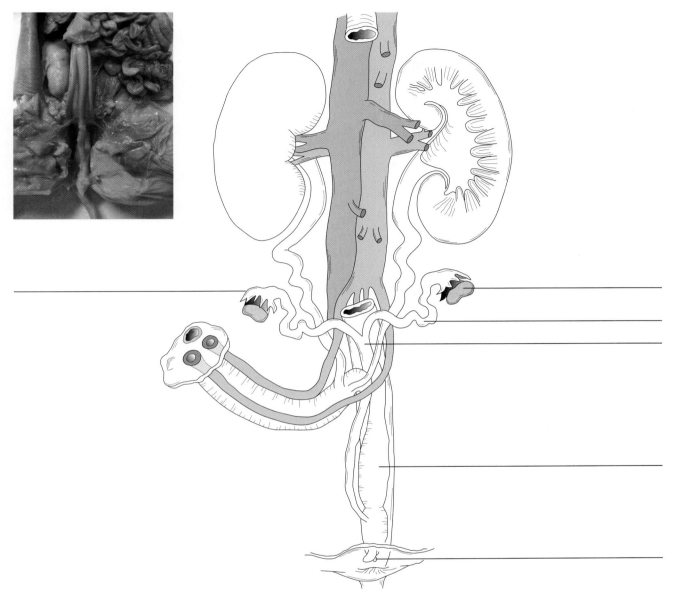

FIGURE 19-2 Reproductive System of the Female Pig

5. Apply your knowledge of the male and female reproductive systems by matching each of the following organs or structures with the appropriate functions below.

Answers can be used **more than once**. Some questions may have **more than one answer**. If so, **include them all**.

A. bulbourethral gland
B. urethra
C. vagina
D. prostate gland
E. fallopian tube
F. epididymis
G. vas deferens
H. urogenital opening
I. uterus
J. ovaries
K. testes
L. penis

____ testosterone is produced here

____ passageway for sperm into the female reproductive tract

____ structure that lies between the ovary and the uterus

____ structure that is part of both the urinary and reproductive systems of pigs

____ fetal development occurs here

____ estrogen and progesterone are produced here

____ sperm production, by meiosis, occurs here

____ produces fluid that transport sperm through the male reproductive tract

____ tube located between the epididymis and the urethra

____ can be closed in a birth control operation for females

Biotechnology Today

A new application of infrared technology is being used to improve the pregnancy rate for women who need in vitro fertilization. The new technique can evaluate the rate of metabolic activity by taking a sample of the fluid that surrounds the embryo. The technique provides a metabolic profile that can be used to identify the best and most viable embryos for implantation.

To develop the technique, researchers tested around 500 samples of embryonic fluid from previous embryo transfers, without knowing which had implanted successfully. Using existing prediction methods, implanted embryos have about a 40% probability of ending in a successful pregnancy, but with the infrared method of selection, the success rate increased by almost 15%. Infrared technology will help doctors to select embryos for transfer that are most likely to result in a successful pregnancy and move toward implanting just one embryo at a time (reducing the chances of multiple births).

ACTIVITY 3 • SIMULATING THE SPREAD OF SEXUALLY TRANSMITTED INFECTIONS

This activity simulates the spread of an infectious disease (chlamydia) through a non-infected population. Chlamydia is the most common sexually transmitted disease in the United States. It's a bacterial infection that causes several serious health complications, including sterility. Approximately four million new infections are reported each year. In this activity, you'll simulate the spread of chlamydia when **one infected person** moves into a neighborhood of sexually active young adults in which no one is infected with the disease.

You'll share the results of your simulation with the class and then calculate the rate of infection, based on the total number of infected people.

INQUIRY AND ANALYSIS

1. Go to the Research Randomizer random number generator at http://www.randomizer.org/form.htm

    Note: If you don't have access to the Internet, **Appendix III** has instructions for using the random number generation feature in MS Excel.

2. Complete the steps to generate **two sets** of random numbers.

   To generate a set of random numbers, enter the choices listed below:

   - How many sets of numbers do you want to generate?   50
   - How many numbers per set?   2
   - Number range   From: 1
       To: 48
   - Do you wish each number in a set to remain unique?   No
   - Do you wish to sort your outputted numbers?   No
   - How do you wish to view your outputted numbers?   Place Markers Off

3. When you've entered your choices, click on **Randomize Now!** The Research Randomizer will generate a set of random numbers for you to print out.

   Staple the random number set and label it **Trial 1.**

   After you have printed out the first set of random numbers, return to the random number generator and click on **Randomize Now!** again.

   Print out the second set of random numbers. **Staple** the second random number set and label it **Trial 2.**

4. Make **four copies** of the simulation grid in **Figure 19-3**. Each sheet will have 12 squares on it. Lay the four sheets next to each other on a flat surface.

5. **Number** the squares from **1–48.**

   Randomly place a **white bean** into **one** of the 48 squares. Place **one dark bean** into **each of the 47** remaining squares.

   The **dark beans** represent **uninfected members** of the population. The **white bean** represents the **infected person,** newly arrived in your neighborhood.

6. Using the set of random numbers for **Trial 1,** simulate a series of sexual encounters in your population.

   Look at the **first set** of two numbers on the random number set for **Trial 1.** If **both people are uninfected** (dark beans), **continue** to the **second set** of numbers.

   If one of the people in the pair **is infected** (white bean), **replace the dark bean** representing the uninfected partner with a **white** bean (the partner is now infected!). Continue on to the next set of numbers.

   If **both** are **already infected, continue** to the next set of numbers without replacing any beans. Continue until you've **completed the 50 interactions for Trial 1.**

   Note: Don't remove the beans from the sheets of printed squares.

FIGURE 19-3 Grid for STD Simulation

7. Record your **Trial 1 results** in **Table 19-1**.

TABLE 19-1 Individual Results for Transmission Simulation

	Number of Dark Beans	Number of White Beans
Start of Simulation	47	1
Trial 1		
Trial 2		

8. Leaving the beans in their current positions, begin **Trial 2**.

   Follow the same procedures you used to complete Trial 1, using the **second set** of random numbers that you generated.

   At the completion of the 50 interactions, **record your Trial 2 results** in **Table 19-1**.

9. Calculate the changes in **infection rate** produced by your simulation, as follows:

$$\text{infection rate} = \left(\frac{\text{number of white beans}}{48}\right) \times 100$$

   Initial infection rate (before simulation)   =   2.1%

   Infection rate after Trial 1   =   _____%

   Infection rate after Trial 2   =   _____%

10. Was there a difference in the infection rate between Trials 1 and 2? If so, suggest one reason why the difference occurred.

11. **Post your experimental results** for Trials 1 and 2 according to the directions given by your instructor.

    At the end of the posting period, **total the results from the class and record them in Table 19-2.**

TABLE 19-2	Class Results for Transmission Simulation	
	Number of Dark Beans	Number of White Beans
Start of Simulation		
Trial 1		
Trial 2		

12. **Calculate** the number of dark and white beans at the start of the simulation as follows:

    Number of dark beans = 47 × number of students who posted results

    Number of white beans = number of students who posted results

    To calculate the number of dark and white beans **after** Trials 1 and 2, simply add the number of dark and white beans recorded in each posting.

13. **Calculate** the changes in infection rate produced by the class simulation (for comparison with your individual results).

$$\text{class infection rate} = \left(\frac{\text{number of white beans}}{48 \times \text{number of students posting results}}\right) \times 100$$

   Initial infection rate (before simulation) = 2.1%

   Class infection rate after Trial 1 = _____%

   Class infection rate after Trial 2 = _____%

14. Were there any differences in infection rate between your individual results and the class results? If so, what were the differences?

15. Based on your understanding of the scientific method, which results do you think would be more reliable: **the class results or your individual results? Explain your answer.**

16. If you randomly designated **four** members of your population to wear condoms for each sexual encounter, how do you think that would affect the rate of transmission? **Explain your answer.**

17. List and explain behaviors that lead to an **increase** in the rate of transmission in the population.

18. List and explain two different actions **other than** condom use that could be taken to **reduce** the rate of transmission.

19. Based on the class simulation results, why are sexually transmitted diseases such as chlamydia, herpes, and HIV so common in the American population?

Exercise 20 • The Immune System

OBJECTIVES

After completing this exercise, you should be able to:

- summarize the functions of various types of physical and chemical barrier defenses of the immune system
- list and explain the components of the inflammatory response and how each is helpful to immune system defense
- summarize the functions of the white blood cells and chemical weapons that are part of the nonspecific defenses
- explain how "self" and "foreign" antigens are related to immune system function
- describe the specificity of antibodies to antigens
- describe how vaccines boost the body's specific defenses
- discuss the benefits of vaccination for public health

SUPPLIES

Activity 1

SUPPLIES FROM LAB KIT

- Pipette, 1

HOUSEHOLD SUPPLIES

petroleum jelly (example: Vaseline®), 1 container

bread, 2 slices

butter, margarine, or other spread

food coloring, 1 drop

vinegar, ½ cup

ground black pepper, 1 teaspoon

antacid tablet, 1

small plate, 2

small bowl (or plastic container), 1

index card, blank, white, 1

ACTIVITY 1 • BARRIER DEFENSES

Do you know someone who's a "germophobe"? If we're surrounded by germs, why aren't we always sick? How does the immune system protect us from infection? The immune system has several layers of protection, beginning with barrier defenses.

In this activity, you'll perform three experiments that demonstrate how the barrier defenses of the immune system function.

Physical Barriers

The skin is the largest organ in the body. The tough epidermis serves as a barrier between you and the outside world. Barriers include skin oils and proteins within the epidermis.

INQUIRY AND ANALYSIS

1. Place two slices of bread on separate small plates. Spread one piece of bread with butter or margarine. Completely cover one side of the bread with the spread.

    Place one drop of food coloring, gently, in the center of each slice of bread. **Observe** the results.

2. What happened to the food coloring on each slice of bread?

    a. without the butter:

    b. with the butter:

3. Which slice of bread was the control? **Explain your answer.**

4. If the slice of bread represents a part of your body that needs protection, what does the butter represent? **Explain your answer.**

5. Does this work in real life? To find out, hold one hand completely level, palm side down. Using the pipette, gently place a drop of water in the center of the back of your hand. Keep your hand level! **Observe and record** your results.

6. How were your results similar to or different from the bread and butter experiment?

Chemical Barriers

Chemical secretions that act as barriers in various locations form part of the body's protective shield.

For example, the surface of the skin, the vagina, and the urethra all have a **low (acidic) pH** that prevents the growth of many microorganisms.

In addition to lowering pH, the body also secretes enzymes that destroy pathogens. For example, the **antibacterial enzyme lysozyme**—found in body fluids such as tears, sweat, and saliva—kills bacteria by destroying their cell walls.

INQUIRY AND ANALYSIS

1. Take one antacid tablet and cut it into several small pieces.

   Place approximately 1/2 cup of vinegar in a small bowl or plastic container. Drop the antacid pieces into the vinegar. Observe the results.

   What happened to the antacid pieces when placed in the vinegar?

2. Household vinegar is approximately pH 2.5. Stomach acid with a pH of 1.5 is much more acidic. Based on the results of your experiment with the vinegar, what do you think would happen to bacteria that enter the stomach with your food?

3. What action of the digestive system was simulated when you broke the antacid tablet into pieces?

Mucous Membranes

The epithelial linings of many tissues are coated with a **sticky mucus** that can trap microorganisms and other particulates. In some locations, such as the trachea and respiratory passageways, **cilia** on the epithelial cells move the trapped particles into the mouth where they're either coughed out or swallowed.

INQUIRY AND ANALYSIS

1. Draw two circles, each two inches in diameter, on a blank, white index card. Spread a thin layer of petroleum jelly in **only one** of the circles.

Place the index card on the counter or on a table. Make a small mound of pepper on the counter a few inches away from the index card. Blow the pepper toward the index card.

 Caution! Don't inhale the pepper.

What does the pepper represent?

2. Leaving the card in position, observe the contents of each circle. **Observe and record** your results.

3. Lift the card and shake it off. Observe the contents of each circle. What was the effect of the petroleum jelly?

4. What body substance is simulated by the petroleum jelly?

5. **Figure 20-1** shows air pollutants trapped in the mucous membranes lining the trachea. **Draw an arrow** on the figure showing the most probable direction the cilia would transport the trapped particulates.

6. If trapped particulates wind up in the mouth and are swallowed, what happens to them?

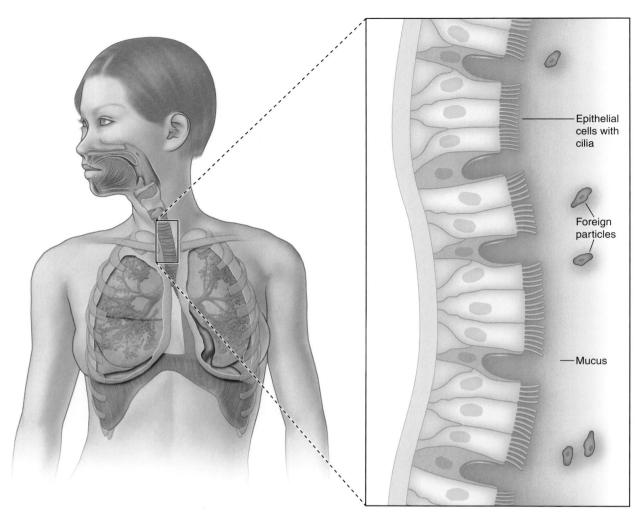

FIGURE 20-1 Mucous Membranes of the Respiratory Tract

7. Nicotine in cigarette smoke paralyzes the cilia in the trachea and other respiratory passages. Considering the results of your simulation, what effect would this have on a smoker's lungs? **Explain your answer.**

ACTIVITY 2 • NONSPECIFIC DEFENSES

If pathogens are able to pass the barrier defenses and enter the body, the **nonspecific defenses** are activated. The nonspecific defenses are carried out by certain types of white blood cells and an array of chemical weapons that manipulate body responses to help fight the infection. For example, **Figure 20-2** shows a type of white blood cell called a macrophage, which seeks out and feeds on foreign bacteria in the bloodstream (a process called phagocytosis).

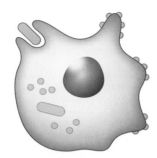

FIGURE 20-2 **Macrophage at Work**

Inflammation is a major component of the nonspecific defenses. You can see the external symptoms of inflammation in an insect bite or a rash (such as poison ivy). They include redness, swelling, heat, and pain. Each of these inflammatory responses helps make the fight against pathogens more effective.

Redness occurs because blood vessels dilate and increase blood flow to the damaged area. Fluid released into the tissues from the bloodstream carries an array of chemicals needed for defense and repair of injuries (and leads to swelling). Pain is the body's way of reminding you to be careful not to further damage the injured area.

A slight increase in body temperature can be very beneficial to the immune system defense. For each 1°C rise in body temperature, metabolic rate increases by 10%. The chemical weapons of the immune system are produced more quickly and work faster at higher temperatures.

All four factors (swelling, redness, increased local temperature, and pain) act in concert to increase the rate of cellular activity in the damaged area and facilitate healing.

Table 20-1 briefly summarizes several mechanisms of the nonspecific defenses.

TABLE 20-1 Nonspecific Defenses

Component	Action
White Blood Cells	
macrophage	phagocytic cell ("big eater") that seeks out and feeds on foreign bacteria in the bloodstream
natural killer (NK) cell	uses chemical weapons (such as perforin) to destroy the cell membranes of cancer cells and cells infected with viruses
mast cell	releases histamine as a signal that triggers inflammation
Chemical Weapons	
perforin	protein that creates holes in the cell membranes of target cells, causing cell death
histamine	signaling protein that enhances the inflammation process by increasing blood flow to the injured area
complement proteins	a diverse group of proteins that enhance immune defenses by: 1) attracting macrophages to the area to engulf pathogens; 2) coating pathogen cell membranes, making them easier for macrophages to destroy; 3) puncturing cell membranes by creating holes similar to the action of perforin; 4) stimulating inflammation by increasing the release of histamine from mast cells
interferons	proteins released by cells infected with viruses that slow the spread of viruses in two ways: 1) attracting macrophages and NK cells to destroy the infected cells, and 2) interferons spread to neighboring cells and stimulate them to produce proteins that prevent further spread of the virus.
pyrogens	circulating proteins that can reset the body's thermostat and raise body temperature (either locally or systemically)
prostaglandins	chemicals that increase blood flow and stimulate nerve endings and send pain signals to the brain

INQUIRY AND ANALYSIS

Using the information in the table, **fill in the blank** with the appropriate immune cells or defensive chemicals. Answers can be used **more than once.**

1. A _____ is an example of a white blood cell that consumes and destroys foreign organisms in the bloodstream.

2. This immune system protein helps slow the spread of a viral infection within your body. _____

3. Anti-inflammatory drugs such as aspirin, acetaminophen, and ibuprofen reduce fever and pain. Therefore, the drugs probably inhibit the production of which two types of immune system chemicals?

   _____ and _____

4. When you cut yourself, _____ cells release _____, which starts the inflammatory response.

5. When you're fighting infection, _____ cells release _____ to destroy the cell membranes of the pathogens.

6. One type of protein that produces a chemical "trail" that macrophages can follow to the site of an infection. _____

7. Tamiflu®, an anti-flu medication, inhibits viral reproduction. Which immune system chemical is it mimicking? _____

8. You're hanging a picture in your house and you accidently hit your thumb with the hammer. Your thumb becomes swollen, red, and painful. As a biology student, explain which immune system actions caused each of your symptoms. **Be specific.**

9. Normal body temperature for a healthy person is approximately 37°C. If your body temperature was 38°C, would it be beneficial to take aspirin or some other medication that reduces fever? **Explain your answer.**

ACTIVITY 3 • ANTIBODY/ANTIGEN SPECIFICITY

If a pathogen gets past the nonspecific defenses, the immune system has one additional response called the **specific defenses.** Body cells are marked with **"self" proteins** that identify them as belonging to you. An **antigen** is a foreign molecule (usually a protein) that activates the immune system defenses. **Antibodies,** produced by specialized B cells in response to antigens, are protein molecules that serve as weapons of the specific defenses.

The specific defenses are highly flexible and responsive. Flexibility is required, since a person can be exposed to millions of different antigens during their lifetime. Each type of antibody differs in shape and structure such that a particular antibody has a shape that matches only one specific antigen.

Consider the following scenario. Ten days ago, you went to a business meeting in a large office building. In the elevator, on the way up, the person next to you, who had a bad cold, sneezed on your coat. Now your body has been flooded with foreign antigens. Antibodies produced by your B cells are on the job, but each antibody is specific. They can only act against the specific antigen they were produced to disable.

In the following simulation, you'll help your antibodies locate and disable the antigens in your bloodstream.

INQUIRY AND ANALYSIS

1. Go to **www.whfreeman.com/bres**

   Click on **Exercise 20** (The Immune System).

   You'll find a **Simulation** created for **Activity 3: Antibody/Antigen Specificity**.

   Follow the onscreen directions to complete the activity and record your answers below.

2. How many antigens could you disable before the time expired?

   Trial 1 results: _____ antigens disabled

3. Try again and see if you can improve your speed and accuracy.

   Trial 2 results: _____ antigens disabled

4. When matching the antibodies to the antigens, what clues did you use to make the correct matches?

5. Did you make any mistakes matching the antibodies to the antigens? What types of errors did you make (for example, visual identification, mouse error, etc.)?

6. After you arrive home from your business trip, you receive a call from the health department notifying you that the person coughing next to you on the plane had tuberculosis. Can your previously formed antibodies protect you against the risk of a tuberculosis infection? Why or why not?

7. Imagine that one of the antigens in the simulation was very similar to the "self" identifiers on your healthy body cells. What immune system malfunction could occur as a result of the similarity? Could the malfunction cause any problems in your body?

ACTIVITY 4 • VACCINATION AS A PUBLIC HEALTH TOOL

As stated by the U.S. Centers for Disease Control and Prevention (CDC), "Vaccines are among the 20th century's most successful and cost-effective public health tools for preventing disease, disability, and death. Not only do they prevent a vaccinated individual from developing a potentially serious disease, vaccines routinely recommended for children also help protect the entire community by reducing the spread of infectious agents." (http://www.cdc.gov/vaccines)

Ease of travel has made crossing national boundaries an everyday occurrence. Over a million people travel internationally each day. Although most infectious diseases are uncommon in the United States, foreign visitors often arrive from countries where many preventable diseases are still prevalent. For this reason, it's critical for all adults and children to be up-to-date with their immunizations.

A **vaccine** stimulates the body to produce defensive antibodies even though you're not sick and you haven't been exposed to the disease. How does this work? Vaccines contain **foreign antigens** that stimulate the specific defenses and the formation of **memory cells.** Memory cells provide long-lasting (active) immunity to the diseases you're vaccinated against.

In this activity, you'll conduct a survey to determine the rate of vaccination in your community.

INQUIRY AND ANALYSIS

1. Survey **20 people** ranging in age from **5 to adult**. Record your results in **Table 20-2**.

    Post the total number of "Yes" responses for each vaccination type in the online classroom as directed by your instructor.

TABLE 20-2 Vaccination Survey Results

varicella - chicken pox
DTP - diphtheria, tetanus, pertussis (whooping cough)
MMR - measles, mumps, rubella (German measles)
influenza - annual flu (this year or last year)

Subjects	Enter **Y** (had vaccine), **N** (didn't have vaccine), or **D** (don't know) for each type of vaccination.				
	Vaccines				
	polio	varicella	DTP	MMR	influenza
1					
2					
3					
4					
5					
6					
7					
8					
9					
10					
11					
12					
13					
14					
15					
16					
17					
18					
19					
20					
Total of "yes" responses					

2. Tabulate the **pooled class results. Record** the total number of "Yes" responses for each vaccination type in **Table 20-3**.

TABLE 20-3 Vaccination Survey - Pooled Class Results

varicella - chicken pox
DTP - diphtheria, tetanus, pertussis (whooping cough)
MMR - measles, mumps, rubella (German measles)
influenza - annual flu (this year or last year)

	Vaccines				
	polio	varicella	DTP	MMR	influenza
Total					
Percent vaccinated					

3. Using the following equation, calculate the percent vaccinated for each category using the formula below. Enter the data in **Table 20-3**.

$$\left(\frac{\text{total number of "Yes" responses}}{\text{\# of students posting data} \times 20}\right) \times 100 = \underline{}\%$$

4. **Create a graph** showing the percent who received each vaccination in your class compared to the results for the United States as a whole (for the U.S. data, see **Table 20-4**).

   Prepare the graph on a computer using the instructions in **Appendix I. Submit** the graph as required by your instructor.

TABLE 20-4 Vaccination Survey - U.S. Results*

varicella - chicken pox
DTP - diphtheria, tetanus, pertussis (whooping cough)
MMR - measles, mumps, rubella (German measles)
influenza - annual flu (this year or last year)

	Vaccines				
	polio	varicella	DTP	MMR	influenza
Percent vaccinated	96.3	96.5	96.0	95.6	12.3

*Results reported are for children entering school (most states require vaccinations before children can enter school). No summary data are available for the adult population.

5. Have you been vaccinated for each of the diseases listed in **Table 20-2**? If not, what are the reasons you haven't received the vaccination(s)?

6. In reference to your graph, what similarities and/or differences can you observe between the **pooled class data** and the **U.S. data**?

7. List and explain two factors that could contribute to the observed differences in vaccination percentages.

8. Why do you think most states require vaccinations for children entering school?

9. If your community has a population of 50,000 people and 10% haven't been vaccinated, how many people are at risk for contracting measles?

10. During the past two years, the CDC has recorded an increasing number of measles cases in the United States. The highest number of measles cases since 1996 was seen in the first few months of 2008 alone. Many of these individuals were children whose parents chose not to have them vaccinated. Highly contagious respiratory diseases, such as whooping cough, accounted for thousands of deaths annually (before a vaccine became available). Infections in the United States have increased more than tenfold since 1980.

Can non-vaccination be considered a public health risk? Should legislation be passed requiring adults to be vaccinated for serious transmissible diseases? **Explain your answer.**

Biotechnology Today

Avian influenza kills about half the people it infects, but luckily, transmission between people is uncommon. However, the potential remains for a global pandemic. A new trend for preventing disease transmission is the production of DNA vaccines. The first DNA-based vaccines are being tested against the H5N1 influenza virus (also known as the bird flu).

DNA vaccines are much faster and easier to make than conventional flu vaccines, which must be grown in eggs. Fragments of the flu viruses' DNA are injected into the body (similar to the injection for a normal vaccine). Once inside the body, the viral DNA triggers the immune system to produce antibodies against the flu virus.

This new approach shows a lot of promise and, in the future, may eventually replace the traditional method of culturing viruses for vaccines.

Appendix I

MAKING A GRAPH IN MICROSOFT WORD 2003

1. Open a blank **Word** document.

2. Press **Enter** three times to create blank lines for text at the top of the page.

3. Click **Insert** from the menu, then click **Picture,** then **Chart.**

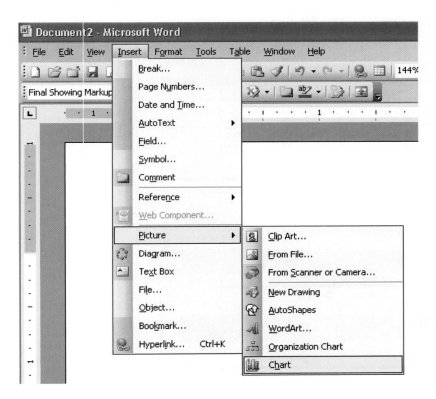

4. Once you do this, a chart and data table will appear within your current document.

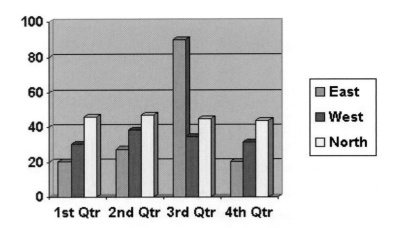

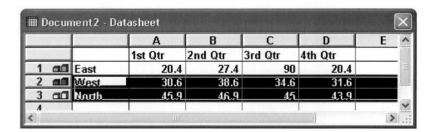

5. Working with the data sheet, right-click the rows of data you will <u>not</u> be using.

6. After highlighting the rows of data, right-click again to make the **Edit menu** appear.

7. Select **delete** to remove the rows of data.

8. At this point, you need to stop and think about how you want to present your data. Using a different type of graph might be a good idea. Three-dimensional graphs are not usually needed for scientific data.

9. Point your cursor toward the current chart until the **Plot Area** box appears.

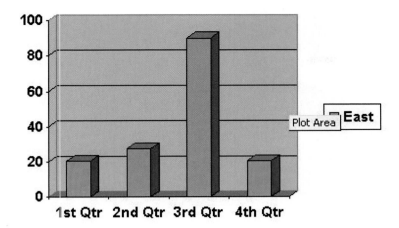

10. Right-click on the current chart and select **Chart Type** from the menu that appears. In the window that appears you can see the variety of chart types you can choose from. Select a chart type, and then close the dialogue box.

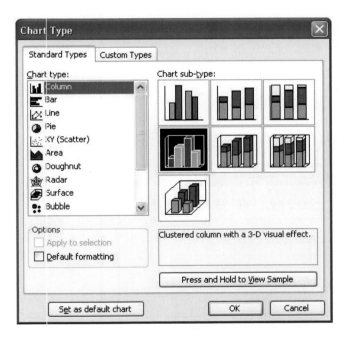

11. Move your cursor to the current graph until the **Chart Area** box appears.

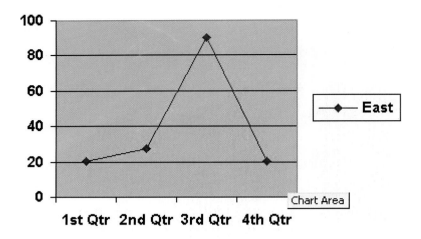

12. Right-click to make the **Format menu** appear and select **Chart Options.**

13. At the top of the **Chart Options** dialogue box, you will see tabs for the various options that are available. Using the available options you can:

   a. Enter titles for the chart, x-axis, and y-axis

   b. Select options for the x and y axes

   c. Make adjustments to the gridlines for the graph

   d. Make changes to the chart Legend, Data Labels, and Data Table

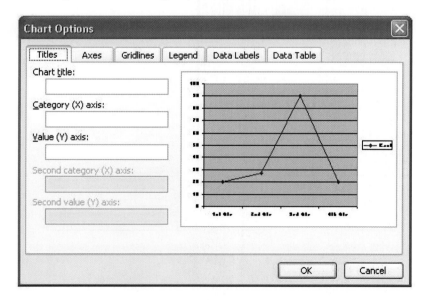

14. Now it's time to enter your data. In cell, **A1** enter your first data point. Enter your second data point in cell **B1**. Continue in this manner until you've entered all of your data in the **row numbered 1.**

		A	B	C	D	E
		1st Qtr	2nd Qtr	3rd Qtr	4th Qtr	
1	East	20.4	27.4	90	20.4	
2						
3						
4						

15. Enter the information that will be on your **X axis** in the row that is not numbered. When you've filled in the data sheet, close the data sheet.

		A	B	C	D	E
		1st Qtr	2nd Qtr	3rd Qtr	4th Qtr	
1	East	20.4	27.4	90	20.4	
2						
3						
4						

Appendix II

STARCH TEST RESULTS

Coleus Leaf Before Iodine Test

Coleus Leaf After Iodine Test

Appendix III

USING THE RANDOM NUMBER GENERATOR IN MICROSOFT EXCEL

Microsoft Excel 2003 or 2007

1. Open a blank **Excel** document.

2. To generate the first set of random numbers, select **Rows 1 through 50** for **columns A and B**.

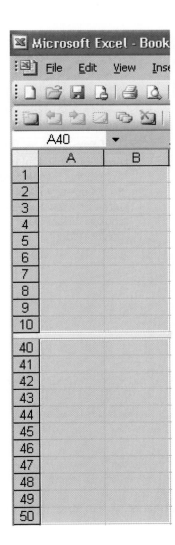

3. Using the formula bar, enter the following text: =randbetween(1,48)

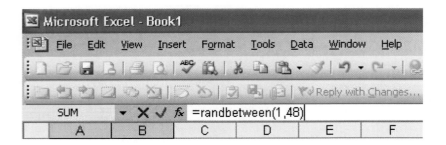

> **Note:** If this function is not available on your computer and you get a, **#NAME? error,** you will have to install and load the Analysis ToolPak add-in. To do this, select **Tools** from the menu, then click on **Add-Ins.**

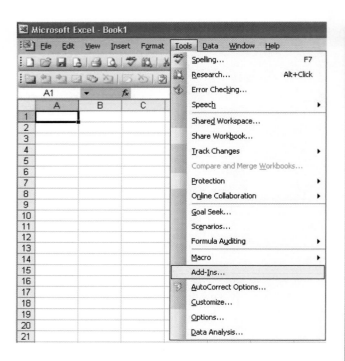 

4. On the dialogue box that appears, check **Analysis ToolPak,** then click **OK.** Excel will then install the add-in on your computer.

5. Click **Ctrl+Enter** to fill all of the highlighted cells with the randomly generated numbers.

6. To make sure that Excel doesn't create a new set of numbers, you have to tell Excel not to recalculate your formula.

   Excel 2007: Select **Formulas** from the menu, then **Calculation Options,** then choose **Manual.**

   Excel 2003: Select **Tools** from the menu, then **Options.**

   In the dialogue box that appears, click on the **Calculations** tabs. Next select **Manual** in the **Calculations** area of the dialogue box. Finally, click **OK** at the bottom of the dialogue box.

8. To generate the second set of random numbers, **repeat instructions 1–7.**

9. **Save your file.** Print the three sets of numbers to use with the STD simulation from the Reproductive System chapter.

Appendix IV

PHOTO INDEX OF YOUR LAB KIT SUPPLIES

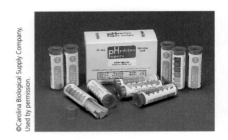

pH Paper

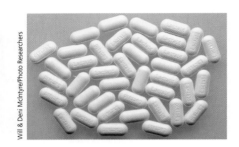

Glucose Strips

Lactaid Tablets

Protein Strips

Pipette

Lime Water, Sealed Tube

Strip Thermometer, Wide Range

Strip Thermometer, Narrow Range

Dialysis Tubing

Stamp Pad

Magnifying Glass